Aerodynamics, Aeronautics, and Flight Mechanics

Aerodynamics, Aeronautics, and Flight Mechanics

Editor

Cezar Dalca

Aerodynamics, Aeronautics, and Flight Mechanics
Edited by **Cezar Dalca**

Printed in 2017

ISBN: 978-1-68117-106-7
Library of Congress Control Number: 2015951988

© 2016 by
SCITUS Academics LLC,
616, Corporate Way, Suite 2, 4766,
Valley Cottage, NY 10989

www.scitusacademics.com

Notice

Preface

Aeronautics is defined as "the science that treats of the operation of aircraft: also, the art or science of operating aircraft." Basically, with aeronautics, one is concerned with predicting and controlling the forces and moments on an aircraft that is traveling through the atmosphere.

A single comprehensive in-depth treatment of both basic and applied modern aerodynamics. The fluid mechanics and aerodynamics of incompressible and compressible flows, with particular attention to the prediction of lift and drag characteristics of airfoils and wings and complete airplane configurations.

Designed for courses in aerodynamics, aeronautics and flight mechanics, this text examines the aerodynamics, propulsion, performance, stability and control of an aircraft. This book captures some of the new technologies and methods that are currently being developed to enable sustainable air transport and space flight. It clearly illustrates the multi-disciplinary character of aerospace engineering, and the fact that the challenges of air transportation and space missions continue to call for the most innovative solutions and daring concepts.

Table of Contents

CHAPTER 1

Aerodynamics of the Cupped Wings During Peregrine Falcon's Diving Flight

Benjamin Ponitz, Michael Triep, Christoph Brücker

Institute of Mechanics and Fluid Dynamics, TU Bergakademie Freiberg, Freiberg, Germany

ABSTRACT

During a dive peregrine falcons can reach velocities of more than 320 km/h and makes themselves the fastest animals in the world. The aerodynamic mechanisms involved are not fully understood yet and the search for a conclusive answer to this fact motivates the three-dimensional (3-D) flow study. Especially the cupped wing configuration which is a unique feature of the wing shape in falcon peregrine dive is our focus herein. In particular, the flow in the gap between the main body and the cupped wing is studied to understand how this flow interacts with the body and to what extend it affects the integral forces of lift and drag. Characteristic shapes of the wings while diving are studied with regard to their aerodynamics using computational fluid dynamics (CFD). The results of the numerical simulations via ICEM CFD and OpenFOAM show predominant flow structures around the body surface and in the wake of the falcon model such as a pair of body vortices and tip vortices. The drag for the cupped wing profile is reduced in relation to the configuration of opened wings (without cupped-like profile) while lift is increased. The purpose of this study is primarily the basic research of the aerodynamic mechanisms during the falcon's diving flight. The results could be important for maintaining good maneuverability at high speeds in the aviation sector.

INTRODUCTION

The peregrine falcon (Falco peregrinus) is one of the world's fastest birds. During horizontal flight, it reaches velocities of up to 150 km/h ([1] [2]) and even more than 320 km/h when nose-diving to attack its bird prey (e.g. [3]–[10]). Nearly all bird species can alter the shape of their wings and thus they can change their aerodynamic properties [11] [12], a concept known as "morphing wing" [13]. During a dive, peregrines also alter the shape of their wings; while accelerating, they move them closer and closer to their body [10]. Several body shapes can be described as a classical diamond shape of the wings followed by a tight vertical tuck with a cupped-like profile of the frontal wing parts [10] [14]–[16] . Only at top velocities (up to at least 320 km/h) peregrines build a wrap dive vacuum pack, i.e. the wings are completely folded against the elongated body [17]. Peregrines are not only extremely fast flyers but also maintain remarkable maneuverability at high speeds. For instance, during courtship behavior they often change their flight path at the end of a dive, i.e. they turn from a vertical dive into a steep climb. This suggests that peregrines are exposed to high mechanical loads.

Although the nose-diving flight of peregrines has been investigated for numerous times, exact numerical flow simulations have not been carried out. Therefore, we investigated different wing configurations (opened wings and cupped wings extension [9]) of the peregrine falcon during diving flight. Both geometries are gained from a previous study by Ponitz et al. (2014) [18] at maximum diving velocities from a dam wall dive. The present study investigates the influence of the geometry change of the extended cupped wings configuration via numerical simulations. The results show that the cupped wings reduce both form and induced drag, while increasing the lift coefficient. This shows how fine the bird can tune the body forces by morphing the wing shape in diving flight conditions.

MATERIALS AND METHODS

Geometries and Models of the Falcon
Dam Wall Diving Flights of Real Falcons
The numerical simulation in this study is based on the specific diving flight condition which is gained from a previous study by [18]. A peregrine falcon was trained to dive from a dam wall and the dive was captured with a stereo high-speed camera system to reconstruct the 3-D flight path. Furthermore, a camera equipped with 400 mm zoom lenses

was used to gain detail studies of the falcon body geometry. The region of interest shows three flight phases: (1) acceleration/diving phase, (2) transient phase with roughly constant speed and (3) deceleration phase. A maximum dive speed of 22.5 m/s was derived from the flight trajectory at flight phase (2). During the dive two significant body geometries are determined around the transient flight phase (2): one configuration without cupped wings (hereinafter called "opened wings") and one configuration with cupped wings [9] (see Figure 1). For the investigated flight situation an angle of attack $\alpha=5°$ was determined for the equilibrium condition at the maximum diving speed of 22.5 m/s. For this flight situation wings remain stationary and do not flap.

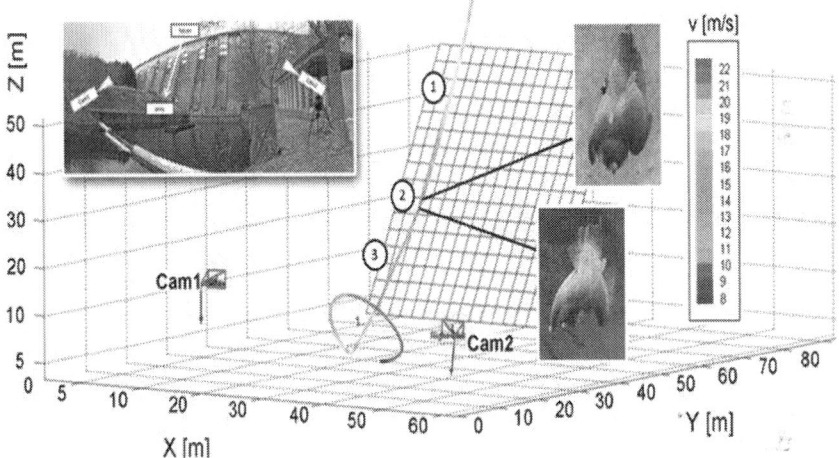

Figure 1. 3-D flight path and detailed body geometries of the falcon's dive from dam wall Olef-Talsperre, Germany. The trajectory is color-coded with the flight velocity magnitude (red color: higher velocities). The region of interest shows three flight phases: (1) acceleration/diving phase, (2) transient phase with roughly constant speed and (3) deceleration phase. Maximum velocity during the dive was 22.5 m/s at flight phase (2). For this transient phase two specific configurations (opened wings and cupped wings) were obtained. An angle of attack α = 5° was determined for the equilibrium condition [18].

Generation of the Models

A life-sized model was built for both configurations, opened wings and cupped wings. Thus, we used the stuffed body of a female peregrine falcon and manually modified its wings until every body shape matched to the geometries of the falcon during dive from the dam wall (see Figure 1). The main part of the body for both configurations is identical. Only the tips of the wings are modified by a kind of winglets extension (cupped wings). Each modified body was fixated and subsequently scanned to acquire its 3-D surface contours (see Figure 2). In a further step, a one-to-one polyvinyl chloride (PVC) model was fabricated by laser sintering process using the acquired 3-D data. The experimental data from wind-tunnel test on this PVC falcon model is needed to evaluate the numerical simulations. Details of the aerodynamic relevant surface areas are given in Table 1.

Experimental Set-Up

For evaluating the numerical model the flow around the life-size PVC model was measured in a type Göttingenwind-tunnel. A force-balance delivered experimental data of the lift and drag forces of the identical model geometry. Hence, results of the wind-tunnel tests can be compared with numerical simulation results of lift and drag coefficients. The experimental set-up is depicted in Figure 3.

Numerical Set-Up

The numerical flow simulations of the falcon allow visualizing flow regions that are difficult to access by experimental methods. Furthermore, they enable to show distributions, for example, of the wall shear stress or pressure in order to identify hot spots, and most important for the present work, the exact determination of quantities like the λ_2 vortex criterion of the flow field. In the present study, the three-dimensional CAD model of the falcon is transferred into a computational unstructured grid using a grid generation tool ICEM CFD 14.5 (ANSYS, Inc., Canonsburg, PA, USA). The computational domain as marked in Figure 4 includes the inflow region, the falcon region, and the downstream wake region of the flow.

Special attention was paid to the meshing of the falcon. Refinements toward near-wall regions were taken into consideration. The grid consists in total of 6.5 million unstructured tetrahedron cells and 1.5 million prism cells on the falcon surface. A mesh independency check for the results of lift and drag coefficients was done for up to 10 million cells. Simulation stability was investigated in respect to different grid parameters and following settings leads to stable results: The height of the first prisms layer on the falcon surface is set to 0.1 mm $(y^+ = 0.2)$ with a growth factor of 1.10 for the following layer perpendicular to the wall and a total number of 10 layers

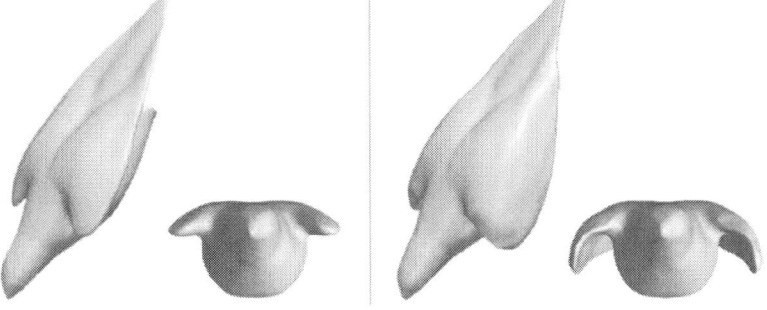

Figure 2. Computer aided design (CAD) falcon models: opened wing shape (left) and cupped wing shape (right).

Table 1. Reference areas of the falcon models with opened wings and cupped wings.

Aspect view	Reference area	
	Opened wings	Cupped wings
Top-view projection area	$A_{ref,opened,top} =$ 0.0411 m^2	$A_{ref,cupped,top} =$ 0.0421 m^2
Frontal projection area	$A_{ref,opened,front} =$ 0.0123 m^2	$A_{ref,cupped,front} =$ 0.0139 m^2

Figure 3. PVC falcon model and force-balance in the wind- tunnel [18] .

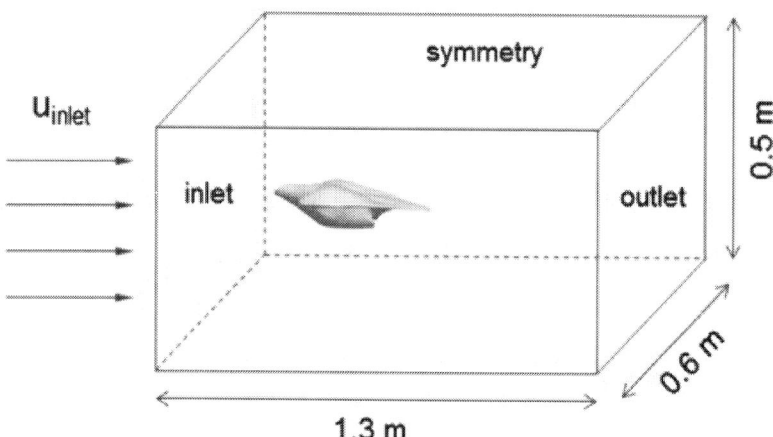

Figure 4. Numerical model dimensions and boundary conditions.

(Figure 5). For these simulation parameters the results deliver stable values which furthermore match the experimental results of lift and drag forces obtained from the wind-tunnel tests (e.g. $C_{D,EXP} = 0.0698$ and $C_{D,CFD} = 0.0725$ for the drag coefficients of cupped wing model geometry). For determining the lift and drag coefficients we used the top-view projection area (see Table 1) as the reference of the falcon model. Thus, the evaluated numerical simulation model is used for the following investigations.

The numerical flow simulation was performed using the open source CFD software OpenFOAM (OpenCFD Ltd., Bracknell, UK). The code numerically solves the conservation equations of mass and momentum by means of a finite volume approach. The Reynolds number Re based on the length $L_{ref} = 400$ mm reaches values of about $Re = 587000$ for $u = 22.5$ m/s. Therefore, turbulent flow is taken into account by a Reynolds averaged approach and the one equation Spalart-Almaras turbulence model. The turbulent viscosity ν_T for the Spalart-Almaras model can be determined by:

$$\tilde{\nu}_T = \sqrt{\frac{3}{2}} u \cdot I \cdot l$$

(1)

where u is the free stream velocity, I is the turbulence intensity ($I = 0.04\%$, from experiments [18]) and l is the length scale ($l = 0.07 L_{ref}$). The comparison with the results gained from the two equations k - ω SST model showed that the Spalart-Almaras simulations delivered more stable results. A major advantage of this model is that it was developed for flow simulations around an airfoil including wake regions and stall phenomena. Air was treated as a single-phase, incompressible$^{(Ma=0.07)}$, isothermal (20°C) Newtonian fluid with constant density (1.189 kg/m³) and viscosity$^{(18.232\times10^{-6}\ Pas)}$. Boundary conditions were chosen in agreement with the experimental situation described in the section before and are defined in Table 2.

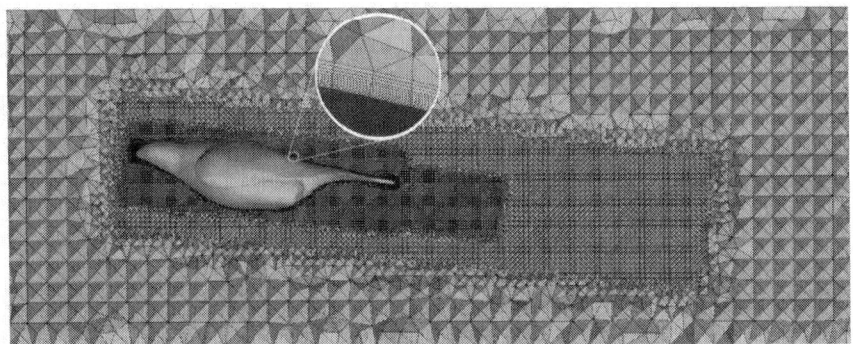

Figure 5. Prisms layer and density boxes of the grid refinement.

Table 2. Overview of boundary conditions.

Boundary	Flow variables		
	u	p	$\tilde{v}_r$
Falcon	$u = 0$	$n \cdot \nabla p = 0$	$\tilde{v}_r = 0$
Inlet	22.5 m/s	$n \cdot \nabla p = 0$	3.08×10^{-4} m²/s
Outlet	$n \cdot \nabla u = 0$	1013×10^3 Pa	$n \cdot \nabla \tilde{v}_r = 0$
Sides	Symmetry	Symmetry	Symmetry

Due to the incompressible character of the fluid, the pressure was set in average constant in the outlet of the test compartment, so that the simulated relative pressure field can be transferred in the post-processing to the correct pressure level with the help of experimental measurements. The simulations were carried out for steady flow conditions.

RESULTS

The angle of attack for the investigated flight situation was determined in a previous study by Ponitz et al. (2014) [18] for the equilibrium condition at the maximum diving speed and is set to $\alpha = 5°$. Figure 6 shows a schematic vector diagram of the acting lift and drag forces on the falcon as well as the angle of attack in relation to the reference line of the falcon model which is built between the falcons tip and tail. The tip of the falcon is defined as the origin of the coordinate system (e.g. $x = 0$ m).

Results from the numerical simulation are post-processed with the software packages ParaView 4.0.1 (Kitware Inc., Clifton Park, NY, USA) and Tecplot 360 2013 (Tecplot Inc., Bellevue, WA, USA). Thus, the integral forces such as lift and drag forces are calculated from the velocity and pressure data field and visualizations are realized. With the calculated lift (L) and drag (D) forces the associated coefficients could be determined as followed:

$$C_L = \frac{L}{q \cdot A} = \frac{2 \cdot L}{\rho \cdot u^2 \cdot A}$$

(2)

And

$$C_D = \frac{D}{q \cdot A} = \frac{2 \cdot D}{\rho \cdot u^2 \cdot A},$$

(3)

where q is the dynamic pressure, ρ is the mass density of the fluid, u is the free stream velocity and A is the reference area of the wing, in both cases the top-view projection area $A_{ref,top}$.

All results are illustrated as comparative data between both configurations, the opened wing shape and the cupped wing shape. Table 3 shows the obtained lift and drag coefficients. Figure 7 shows the visualizations

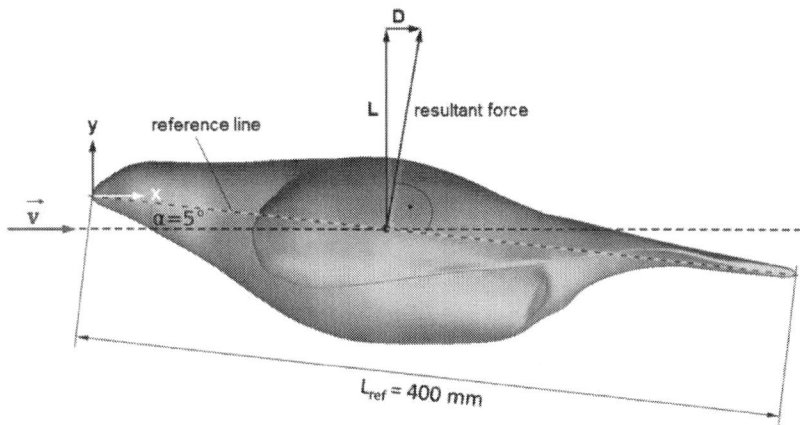

Figure 6. Definition of reference line, angle of attack and schematic vector diagram of forces.

Table 3. Comparison of lift and drag coefficients for both body geometries: opened wings vs. cupped wings (Reference area is the top-view projection area).

Flow parameter	Falcon model configuration	
	Opened wings	Cupped wings
Lift coefficient C_L	0.0851	0.1119
Drag coefficent C_D	0.0892	0.0725

Between the two configurations for surface pressure distribution, wall shear stress, streamlines below the wings, wake flow via λ_2 vortex detection criterion [19] and slices in the wake [20] . In addition, Figure 8 gives a more detailed insight of the surface streamlines on both body configurations.

The comparison of the lift coefficients (Table 3) shows a value of $C_{L,opened} = 0.0851$ for the configuration with opened wings and a value of $C_{L,cupped} = 0.1119$ for the cupped wing shape. The cupped wing profile causes an increased lift coefficient for identical simulation parameters. This is due to the cupped wings which deliver more wing area for lift. The passing air flow is held below the wing on the pressure side of the cupped wing. Hence, the acting lift component is increased in relation to the geometry of opened wing configuration. More interesting is the focus on the drag coefficient. Due to the extended cupped wing shape an increased drag coefficient is expected in relation to the opened wing geometry, however the results show the contrarian. With the cupped wings extension the drag coefficient is reduced from $C_{D,opened} = 0.0892$ down to a value of $C_{D,cupped} = 0.0725$, see Table 3. The reason could be found in the flow details as discussed below by means of streamlines and near-surface flow visualizations.

For comparison with bluff body aerodynamics we calculated in addition the drag coefficient with the projection area along the axis of flight (frontal projection area $A_{ref,front}$). The following values were determined: $C_{D,opened,frontal} = 0.2980$ and $C_{D,cupped,frontal} = 0.2402$. They show the same tendency of decreased drag in the cupped wing configuration.

In the upper part of Figure 7 surface pressure distributions show typical areas of higher pressure on the tip of the falcon body and on the leading edge of the wings. Typical areas of lower pressure are found on the suction side of both wing configuration and especially between the cupped wings and the falcon body. In addition, wall shear stress distributions combined with the Line-Integral-Convolution (LIC) visualization method [21] show characteristic patterns of a streamlined body. Besides the visualizations of surface pressure and wall shear stress distributions, the streamlines and wake flow characteristics below and behind the falcon models show significant flow features (see middle and lower part of Figure 7). For instance, the comparison of the streamlines below the wings (color coded with the velocity magnitude) show different pathways and various hotspots of maximum velocity values below the wings. Streamlines around the opened wing configuration indicates flow separation whereas the cupped wings show streamlines which lies close to the body contour for a much longer distance downstream. Additionally, the spot of higher velocity magnitude below the cupped wings indicate an acceleration of the flow in this region like in a tapering nozzle. Therefore the flow remains attached to the body for a longer distance downstream, which affects the body drag in a beneficial way. This could be one reason for the reduced drag of the cupped wing geometry in relation to the opened wing shape.

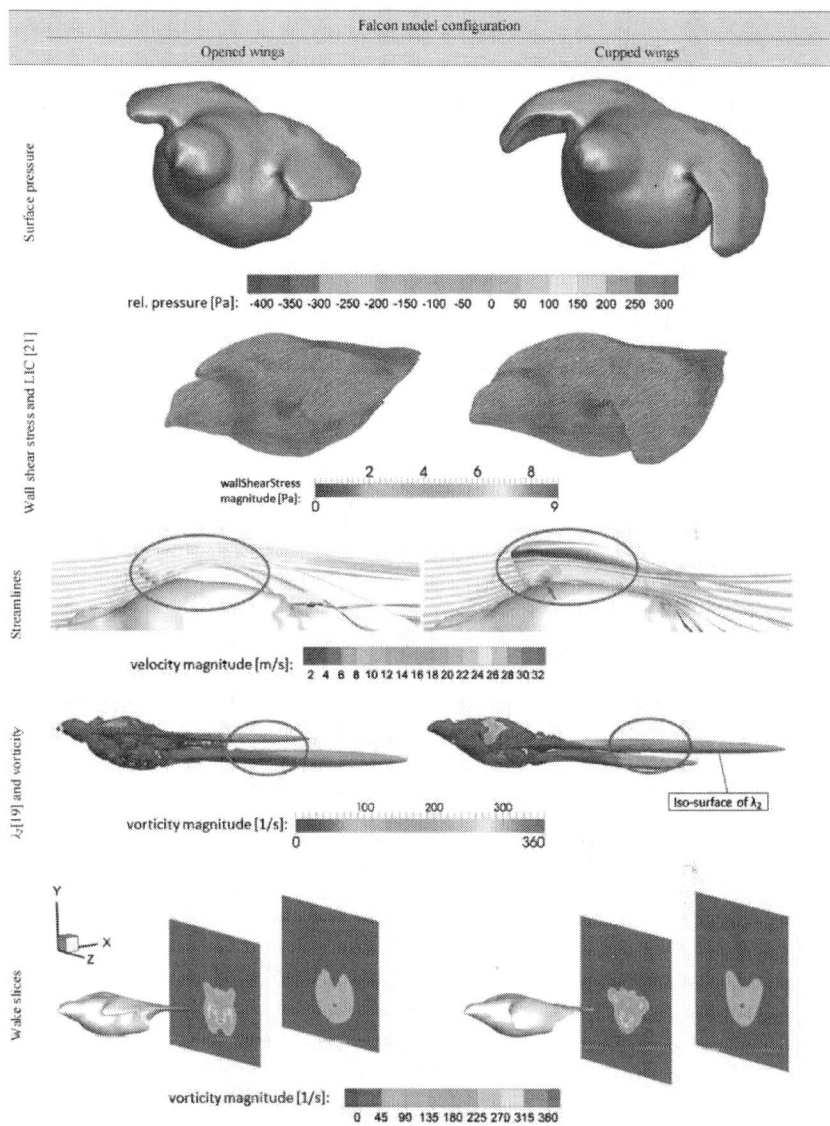

Figure 7. Comparison of visualized flow parameters for both configurations: opened wings vs. cupped wings.

Furthermore, the wake flow behind the model visualized by the λ_2 vortex detection criterion (iso-surface is color coded with the vorticity magnitude) let recognize some differences between both configurations. In both cases two vortex pairs are recognizable. One vortex pair is generated around the wings (known as the wing-tip vortex) and one vortex pair is generated from flow separation at the aft part of the falcon main body (called herein the body vortex). The vorticity distribution in the wake is shown for two discrete slices in the Trefftz- plane [20] at the positions close to $(x=0.5\ \text{m})$ and far behind $(x=0.85\ \text{m})$ the falcon model. The cupped wings lead to wing-tip vortices which are located further down in vertical direction than in the case of the opened wing configuration. Body vortices of both model shapes occur rather in the same location. Hence, the spatial arrangement of the vortex pairs (wing-tip and body vortices) is significantly closer in the case of the cupped wings geometry. In general, the induced drag of a wing depends on the strength of circulation and the lateral position of the wake vortices away from the centerline, thus it is concluded that the observed differences also influence the induced drag for both geometries.

Figure 8 depicts details of the surface streamlines around both body configurations. Streamlines below the opened wings indicates more clearly local flow separation in relation to the geometry with cupped wings where streamlines appear more continuous such as in an attached flow.

Figure 9 illustrates the three-dimensional arrangement of wake vortices visualized by the iso-surface of the vortex detection criterion λ_2 colored coded with the vorticity magnitude. The velocity distribution in the slice at position $x = 0.64\ \text{m}$ shows the downwash in the wake region of the falcon.

DISCUSSION AND CONCLUSIONS

This study investigated in detail the aerodynamics of steady flight conditions of a peregrine falcon in dive motion. The contours of body and wing shape as well as the dive speed and angle of attack have been detected in a previous study and were used herein for numerical flow simulations around the body and in the wake: we simulated the flow at an angle of attack of 5° and a flow speed of 22.5 m/s.

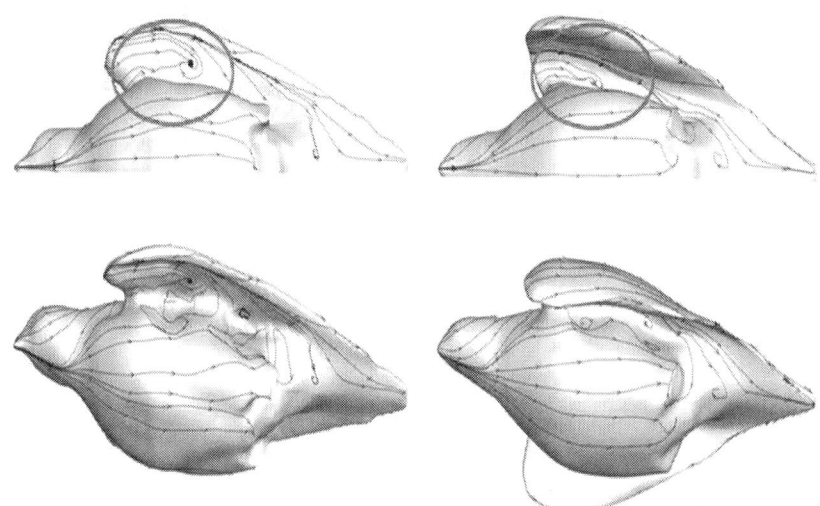

Figure 8. Comparison of visualized surface streamlines for both configurations: opened wings (left) vs. cupped wings (right).

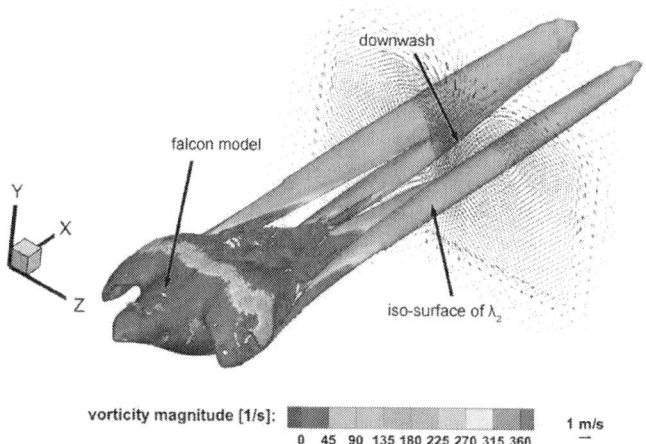

Figure 9. Three-dimensional visualization of the spatial arrangement of wake vortices. The iso-surface of λ_2 criterion is color coded with the vorticity magnitude.

The focus of this study was the comparison of the opened wing shape and geometry with cupped extension of the wing tip. The formulation "cupped wings" was first used by [9] to describe the shape of the downward tilted tips of the wing which is a typical falcon shape during dive. When comparing both wing configurations the results clearly reveal that the cupped wing configuration increases lift and decreases drag under the same flow conditions (and angle of attack). Although the total surface area in the cupped wing configuration is larger, the body drag is reduced. The reason therefore is the acceleration of the flow close to the body in the gap between cupped wing tip and main body surface. Therefore flow separation is shifted towards the trailing edge of the body which reduces the form drag of the body. In addition, the modification of the tip vortices position and strength in the wake of the cupped wings hints on the beneficial effect on reducing the induced drag. This can be deduced from the analysis of the flow in the Trefftz-plane behind the body. This shows how fine the bird can tune the body forces by morphing the wing shape in diving flight conditions.

Conclusions drawn herein are based on a smooth surface of the model. In nature the body is covered with feathers which may also play a role [22] . The tiny scales and the elastic properties of the feathers were not taken into account. This is subject of ongoing work.

ACKNOWLEDGEMENTS

The study is funded by the Deutsche Forschungsgemeinschaft (BL 242/19-1; BR 1494/21-1). The funders had no role in study design, data collection and analysis, decision to publish, or preparation of the manuscript. We thank Horst Bleckmann for his support in gathering the data of the cupped wing model.

REFERENCES

1. del Hoyo, J., Elliott, A., Sargatal, J. and Collar, N.J. (1999) Handbook of the Birds of the World. Vol. 5, Lynx Edicions, Barcelona.
2. Podbregar, N. (2013) Das Geheimnis des Fliegens—Tierischen Flugkünstlern auf der Spur: Strategien der Evolution. Springer, Berlin and Heidelberg, 227-243.
3. Tucker, V.A. and Parrott, G.C. (1970) Aerodynamics of Gliding Flight in a Falcon and Other Birds. Journal of Experimental Biology, 52, 345-367.
4. Orton, D.A. (1975) The Speed of a Peregrine's Dive. The Field, 588-590.
5. Brown, L.A. (1976) British Birds of Prey. Collins, London.
6. Alerstam, T. (1987) Radar Observations of the Stoop of the Peregrine Falcon Falco Peregrinus and the Goshawk Accipiter Gentilis. Ibis, 129, 267-273.http://dx.doi.org/10.1111/j.1474-919X.1987.tb03207.x
7. Savage, C. (1992) Peregrine Falcons. Sierra Club, San Francisco.

8. Clark, W.S. (1995) How Fast Is the Fastest Bird? WildBird, 9, 42-43.

9. Tucker, V.A. (1998) Gliding Flight: Speed and Acceleration of Ideal Falcons during Diving and Pull Out. Journal of Experimental Biology, 201, 403-414.

10. Franklin, D.C. (1999) Evidence of Disarray amongst Granivorous Bird Assemblages in the Savannas of Northern Australia, a Region of Sparse Human Settlement. Biological Conservation, 90, 53-68. http://dx.doi.org/ 10.1016/S0006-3207(99)00010-5

11. Nachtigall, W. (1975) Vogelflügel und Gleitflug Einführung in die aerodynamische Betrachtungsweise des Flügels. Journal für Ornithologie, 116, 1-38.http://dx.doi.org/10.1007/BF01643073

12. Nachtigall, W. (1998) Der Gleitflug von Vögeln. Physik in unserer Zeit, 1, 25-29.

13. Lentink, D., Müller, U.K., Stamhuis, E.J., de Kat, R., van Gestel, W., Veldhuis, L.L.M., et al. (2007) How Swifts Control Their Glide Performance with Morphing Wings. Nature, 446, 1082-1085. http://dx.doi.org/ 10.1038/nature05733

14. Ratcliffe, D.A. (1980) The Peregrine Falcon. Buteo Books, Vermillion.

15. Hustler, K. (1983) Breeding Biology of the Peregrine Falcon in Zimbabwe. Ostrich, 54, 161-171. http://dx.doi.org/10.1080/00306525.1983.9634466

16. Tucker, V.A. (1990) Body Drag, Feathers Drag and Interference Drag of the Mounting Strut in a Peregrine Falcon, Falco peregrinus. Journal of Experimental Biology, 149, 449-468.

17. Seitz, K. (1999) Vertical Flight. NAFA Journal, 38, 68-72.

18. Ponitz, B., Schmitz, A., Fischer, D., Bleckmann, H. and Brücker, C. (2014) Diving-Flight Aerodynamics of a Peregrine Falcon (Falco peregrinus). PLoS ONE, 9, e86506.http://dx.doi.org/10.1371/journal.pone.0086506

19. Jeong, J. and Hussain, F. (1995) On the Identification of a Vortex. Journal of Fluid Mechanics, 285, 69-94. http://dx.doi.org/10.1017/S0022112095000462

20. Krasny, R. (1987) Computation of Vortex Sheet Roll-Up in the Trefftz Plane. Journal of Fluid Mechanics, 184, 123- 155. http://dx.doi.org/10.1017/ S0022112087002830

21. Cabral, B. and Leedom, L.C. (1993) Imaging Vector Fields Using Line Integral Convolution. In: Proceedings of ACM SIGGRAPH'93, Anaheim, 2-6 August 1993, 263-270.

22. Schmitz, A., Ponitz, B., Brücker, C., Schmitz, H., Herweg, J. and Bleckmann, H. (2014) Morphological Properties of the Last Primaries, the Tail Feathers, and the Alulae of Accipiter nisus, Columba livia, Falco peregrinus, and Falco tinnunculus. Journal of Morphology, Early View. http://dx.doi.org/ 10.1002/jmor.20317

CITATION

Ponitz, B. , Triep, M. and Brücker, C. (2014) Aerodynamics of the Cupped Wings during Peregrine Falcon's Diving Flight. Open Journal of Fluid Dynamics, 4, 363-372. doi: 10.4236/ojfd.2014.44027.

CHAPTER 2

Flight Dynamics Modeling of a Small Ducted Fan Aerial Vehicle Based on Parameter Identification

Zhengjie Wang[1], Zhijun Liu[1],Ningjun Fan[1], Meifang Guo[2]

[1] School of Mechatronical Engineering, Beijing Institute of Technology, Beijing 100081, China
[2] North Institute for Science & Technical Information, Beijing 100089, China

ABSTRACT

This paper presents a simple and useful modeling method to acquire a dynamics model of an aerial vehicle containing unknown parameters using mechanism modeling, and then to design different identification experiments to identify the parameters based on the sources and features of its unknown parameters. Based on the mathematical model of the aerial vehicle acquired by modeling and identification, a design for the structural parameters of the attitude control system is carried out, and the results of the attitude control flaps are verified by simulation experiments and flight tests of the aerial vehicle. Results of the mathematical simulation and flight tests show that the mathematical model acquired using parameter identification is comparatively accurate and of clear mechanics, and can be used as the reference and basis for the structural design.

INTRODUCTION

The ducted fan aerial vehicle is a new kind of single-blade rotor aircraft which is different from traditional fixed-wing craft and helicopters. This new kind of aerial vehicle has more compact architecture and the attitude controls can deal with more complex flow patterns. From the beginning of the 1990s, many countries have started research in this field one after another, and have developed different ducted aerial vehicles.[1 and 2] Single-blade ducted aerial vehicles have shown great potential in the military field and humanitarian relief. In recent years, related research institutions have continuously invested money in technology research in different fields, among which aerodynamic configuration design, airframe structure design, testing methods and control methods have become the important fields.[3, 4, 5 and 6].

Ducted fan aerial vehicles are unstable static system. Good controllability is a precondition for the stable flight of an aerial vehicle, and the controllability of the aerial vehicle is closely related to the structural parameters of the vehicle. The precondition for realizing a stable flight is to find proper controllability conditions for the miniature aerial vehicles, and based on those conditions to carry out a structural design for the aerial vehicle. Therefore, it is crucial to find a modeling method which can reflect the features of the internal structure of the vehicle and has comparatively accurate model parameters.

There are two ways to construct a mathematical model for a system. One is mechanism modeling and the other is system identification modeling. Mechanism modeling can produce a mathematical model for a ducted aerial vehicle. Based on this traditional methods, Marconi, et al.[7] studied a simple ducted fan micro air vehicle (MAV) based on a single fixed pitch rotor and four active aerodynamic surfaces and investigated the dynamical model of this MAV. Naldi, et al.[8] provided a nonlinear dynamical model of the system which is able to describe stationary flight, fast forward flight and the transitions between the two flight conditions. Ko, et al.[9] obtained the model of the ducted fan vehicles by a kind of modeling software tool that has been validated through various wind tunnel tests and flight tests of ducted fan vehicles. However, as the architecture of such an aerial vehicle is special, the parameters of the model cannot be obtained accurately using analysis. Using the actual input and output data of the aerial vehicle, the system identification can produce a mathematical model for the aerial vehicle. In contrast, this paper aims to present few preliminary results for acquiring the dynamics model by using experimental methods. Then, based on the obtained dynamics model, the method for the structure design of the control surfaces is carried out.

FLIGHT DYNAMICS MODELING

System composition of ducted fan aerial vehicle

This kind of aerial vehicle is composed of three parts, which are the power system, the attitude control system and the airframe system, as shown in Fig. 1. The power system includes the brushless DC motor, the electronic regulator, the batteries and the ducted propeller. The attitude control system is composed of two layers of aerodynamic control flaps under the propeller. These two sets of control flaps are located within the rotational flow field produced by the propeller. The airframe system includes the ring duct, the equipment cabin and the supporting structure of the airframe. The ring duct can isolate the high-speed spinning ducted propeller from the external environment in order to provide higher security during the operation. Meanwhile, the specially designed duct has significant augmented lift effect.

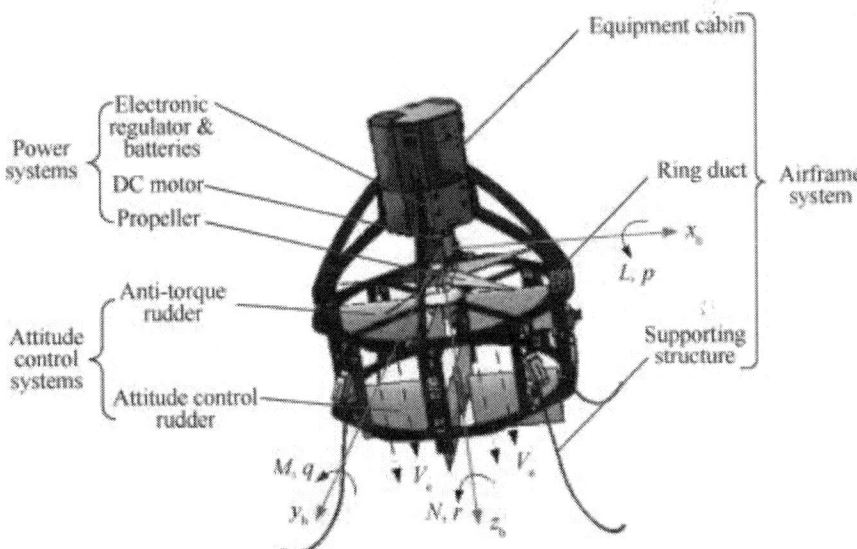

Figure 1. Structural composition diagram.

Flight dynamics modeling

The ducted fan aerial vehicle is a special surface symmetry system. The origin of the coordinate frame set in this paper overlaps with the location of the gravity center of the aerial vehicle. The symmetry planes of the aerial vehicle are $x_bO_bz_b$ and $y_bO_bz_b$, which are shown in Fig. 1 Therefore, the inertia moment of the aerial vehicle is

$$I_{xy}=I_{xz}=I_{yz}=0 \qquad (1)$$

The scalar forms of the corresponding forces and moments are given, respectively, by

$$\begin{cases} F_x = m(\dot{u} + qw - rv) \\ F_y = m(\dot{v} + ru - pw) \\ F_z = m(\dot{w} + pv - qu) \end{cases} \qquad (2)$$

$$\begin{cases} L = I_x\dot{p} + qr(I_z - I_y) \\ M = I_y\dot{q} + rp(I_x - I_z) \\ N = I_z\dot{r} + pq(I_y - I_x) \end{cases} \qquad (3)$$

where F_x, F_y and F_z are aerodynamic forces in $O_b x_b$ axis, $O_b y_b$ axis and $O_b z_b$ axis respectively; L, M and N are rolling moment, pitching moment and yawing moment; u, v and w are axial velocity, lateral velocity and normal velocity; $\dot{u}, \dot{v}$ and $\dot{w}$ are rate of axial velocity, rate of lateral velocity and rate of normal velocity; p, q and r are roll rate, pitch rate and yaw rate; $\dot{p}, \dot{q}$ and $\dot{r}$ are roll, pitching and yawing angular accelerations; m is the mass.

The dynamics equation of the aerial vehicle can be rewritten in vector form, shown as

$$f^b = \begin{bmatrix} m(\dot{u} + qw - rv) \\ m(\dot{v} + ru - pw) \\ m(\dot{w} + pv - qu) \end{bmatrix} \qquad (4)$$

$$\tau^b = \begin{bmatrix} I_x\dot{p} + qr(I_z - I_y) \\ I_y\dot{q} + rp(I_x - I_z) \\ I_z\dot{r} + pq(I_y - I_x) \end{bmatrix} \qquad (5)$$

where f^b and τ^b are vectors of aerodynamic force and moment in body axes.

Parameterization of dynamics equation

In the case of hovering or low speed and level flight, the main forces of the ducted fan aerial vehicle include gravity, thrust and moment caused by the power system, the controlling force and moment caused by the control flaps, and the spinning moment caused by a high-speed spinning propeller. 10 and 11 Therefore, the resultant force and moment of the aerial vehicle are given, respectively, by

$$\begin{cases} f^b = f_g + f_r + f_{flap} \\ \tau^b = \tau_r + \tau_{flap} + \tau_{gero} \end{cases}$$

(6)

where f_g, f_r and f_{flap} are gravity , thrust and control force; and τ_r, τ_{flap} and τ_{gero} are thrust moment , control moment and spinning moment.

The explanation of the relative parameters is given below:

(1) Gravity
The gravitational force acting on the airplane acts at the center of mass (the origin point O_b of the body axis), and its direction is vertical downward. Because the body axis system is fixed at the center of gravity, the gravitational force will not produce any moments. Therefore, we must transform the gravitational force into its body-frame components to give:

$$f_g = \begin{bmatrix} -mg\sin\theta & mg\cos\theta\sin\phi & mg\cos\theta\cos\phi \end{bmatrix}^T$$

(7)

where θ is pitch angle, and ϕ is roll angle.

(2) Thrust and torque of the dynamic system
The thrust of the ducted fan aerial vehicle generated by the propeller has the same direction as the Oz_b axis. Thus the components of thrust corresponding to the Ox_b and Oy_b axes will be zero. The thrust of the dynamic system in the body axis can be written

$$f_r = \begin{bmatrix} 0 & 0 & -K_f n^2 \end{bmatrix}^T$$

(8)

where n is rotating speed, and aerodynamic coefficient K_f produced by thrust f_r is the parameter which needs to be identified.

When the propeller spins about the Oz_b axis at high speed, a moment which makes the vehicle rotate about the Oz_b axis will be produced:

$$\tau_r = \begin{bmatrix} 0 & 0 & K_t n^2 \end{bmatrix}^T$$

(9)

where the aerodynamic coefficients K_t produced by torque τ_r is the parameter which needs to be identified.

(3) Aerodynamic forces and moments of the anti-torque control flaps and the attitude control flaps
Because the anti-torque control flaps use the axial symmetry system, the aerodynamic forces will only generate the resultant force corresponding with the $O_b z_b$ axis:

$$
\boldsymbol{f}_a = \begin{bmatrix} 0 \\ 0 \\ 4\rho S_a V_e^2 (C_{Da}\delta_a^2 + C_{Da0}) \end{bmatrix}
\tag{10}
$$

where ρ the density of the air, S_a the effective lift area of one anti-torque flap, V_e the relative wind velocity generated by the propeller, C_{Da} the drag coefficient for a term quadratic in δ_a, δ_a the angle of the deflection of the flaps, and C_{Da0} the drag coefficient with zero angle of the deflection.

The aerodynamic forces formed by the reaction torque control flaps would produce a moment. This moment rotates about the $O z_b$ axis and defined by

$$
\boldsymbol{\tau}_a = \begin{bmatrix} 0 \\ 0 \\ -4\rho S_a V_e^2 (C_{La}\delta_a + C_{La0})d_a \end{bmatrix}
\tag{11}
$$

where C_{La0} is the lift coefficients with zero angle of the deflection, C_{La} the lift coefficient considering a linear dependency of lift on the angular deflection δ_a, and d_a the distance from the location of the gravity center to the plane which is decided by the eight aerodynamic centers.

The attitude control flaps install in the interior of the ducted fan aerial vehicle, and they are positively under the reaction torque control flaps. The airfoil is the symmetry one NACA0012,[12 and 13] and the corresponding aerodynamic forces and moments are given by

$$
\boldsymbol{f}_c = \begin{bmatrix} \rho S_c V_e^2 C_{Lc}\delta_{c2} \\ -\rho S_c V_e^2 C_{Lc}\delta_{c1} \\ \rho S_c V_e^2 [C_{Dc}(\delta_{c1}^2 + \delta_{c2}^2) + C_{Dc0}] \end{bmatrix}
\tag{12}
$$

$$\tau_c = \begin{bmatrix} \rho S_c V_e^2 C_{Lc} \delta_{c2} d_c \\ \rho S_c V_e^2 C_{Lc} \delta_{c1} d_c \\ 0 \end{bmatrix} \qquad (13)$$

where S_c is the effective lift area of one attitude control flap, C_{Dc} is drag coefficient, C_{Dc0} is zero lift drag coefficient, δ_{c1} and δ_{c2} are the angles of the deflections of the pitch control flaps and roll control flaps respectively, C_{Lc} is lift coefficient, and d_c is the distance between the aerodynamic center of the control flap and the gravity center of the aerial vehicle.

Therefore, the aerodynamic forces and moments caused by the attitude control flaps of the small ducted fan aerial vehicle can be written as

$$f_{\text{flap}} = f_a + f_c = \begin{bmatrix} \rho S_c V_e^2 C_{Lc} \delta_{c2} \\ -\rho S_c V_e^2 C_{Lc} \delta_{c1} \\ f_1 + f_2 \end{bmatrix} \qquad (14)$$

$$\tau_{\text{flap}} = \tau_a + \tau_c = \begin{bmatrix} \rho S_c V_e^2 C_{Lc} \delta_{c2} d_c \\ \rho S_c V_e^2 C_{Lc} \delta_{c1} d_c \\ -4\rho S_a V_e^2 (C_{La} \delta_a + C_{La0}) d_a \end{bmatrix} \qquad (15)$$

Where

$$\begin{cases} f_1 = 4\rho S_a V_e^2 (C_{Da} \delta_a^2 + C_{Da0}) \\ f_2 = \rho S_c V_e^2 [C_{Dc} (\delta_{c1}^2 + \delta_{c2}^2) + C_{Dc0}] \end{cases} \qquad (16)$$

(4) Influence of the spinning moment
The high-speed spinning propeller provides the power for the small ducted fan aerial vehicle. When the attitudes of the vehicle change, the effect of the spinning moment will lead to the coupling phenomenon between the pitch channel and roll channel. The spinning moment can be expressed as

$$\tau_{\text{gero}} = [-I_p \omega_e q \quad I_p \omega_e p \quad 0]^T \qquad (17)$$

where I_p is the inertia moment of the propeller with high-speed, ω_e is the angular rate of the propeller.

Parameter analysis and classification

In accordance with the features and attributes of the parameters, all parameters involved in the above equations are classified. The results of the classification are shown in Table 1.

Table 1. Parameters in mathematical model.

Parameter type	Parameters included	Evaluation method
Structure	$m, I_x, I_y, I_z, I_p, d_c, d_a$	Design and identification
Aerodynamic	$K_f, K_t, C_{La}, C_{La0}, C_{Da}, C_{Da0}, C_{Lc}, C_{Dc}, C_{Dc0}$	Identification
Motion	θ, ϕ, p, q, r , $\dot{p}, \dot{q}, \dot{r}$, u, v, w , $\dot{u}, \dot{v}, \dot{w}$	Flight measurement
Controlling	$\delta_e, \delta_a, \delta_{c1}, \delta_{c2}$	Controlled by input

The structural parameters mainly refer to the mass of the aerial vehicle, the rotational inertia and relative positions of the functional components. These parameters are basically confirmed during the design, and only need to be identified and corrected by experiments during the modeling. Aerodynamic parameters mainly include the aerodynamic coefficients related to the dynamic system, the reaction torque control flaps and the attitude control flaps. These parameters should be identified by experiment. Both motion parameters and controlling parameters are values involved during the actual flight tests, and can be obtained by using airborne sensors and recording devices.

IDENTIFICATION OF STRUCTURAL AND AERODYNAMIC PARAMETERS

Identification of structural parameters

Mass and position parameters

The mass of an aerial vehicle refers to the total mass of all components, including the vehicle structure, the power equipment, the control system and the batteries. Component masses and the total mass of the ducted fan aerial vehicle studied by this paper are shown in Table 2.

Table 2. Mass distribution of systems.

System name	Content	Mass (kg)
Structure of the aerial vehicle	Airframe supporting the structure of the vehicle, undercarriage and other structures.	0.794
Flight control system	Attitude measurement unit, circuit board of the automatic pilot, connecting lines.	0.261
Attitude control system	Anti-torque flaps and pitching and rolling control flaps.	0.215
Dynamic system	Motor, propeller, electron speed regulator.	0.654
Airborne equipment	Cameras and wireless communication module.	0.122
Power source	Power batteries and cables.	0.764
Total mass	Total take-off mass.	2.81

Ducted fan aerial vehicles are axial symmetry systems. In this paper, the center position of the aerial vehicle is confirmed by a bifilar pendulum balance method and adjusted by adjusting the positions of the devices in the cabin. The origin of the coordinate frame of the ducted fan aerial vehicle is set at the position of the center of gravity.

In Fig. 2, the length of the arm of the reaction torque flap for the aerial vehicle equals the distance from the aerodynamic center to the axis $O_b z_b$ of the coordinate frame of the airframe, denoted by d_a. The lengths of the arms of the pitching and rolling control-flaps equals the distances from the aerodynamic center to the axis $O_b y_b$ and axis $O_b x_b$ respectively, denoted by d_c. Here d_a and d_c are design values, where $d_a = 0.08$ m, $d_c = 0.2$ m.

Rotational inertia
Ducted fan aerial vehicles are axial symmetry systems whose origins of the coordinate frame of the airframe are set at the position of the center of gravity. Its inertia moment is $I_{xy} = I_{xz} = I_{yz} = 0$. Therefore, the identification only involves the main diagonal of the inertial tensor. That is the identification of the rotational inertia. During the experiment, the tri-wire pendulum method is used to measure the rotational inertia of the rigid body. The tri-wire pendulum method diagram is shown in Fig. 3. [14]

According to the law of conservation of energy and the law of rotation of rigid body, the rotational inertia of the object rotating around the central axis OO' can be derived as follows:

$$I = \frac{mgR_2R_1}{4\pi^2 H} T_0^2$$

(18)

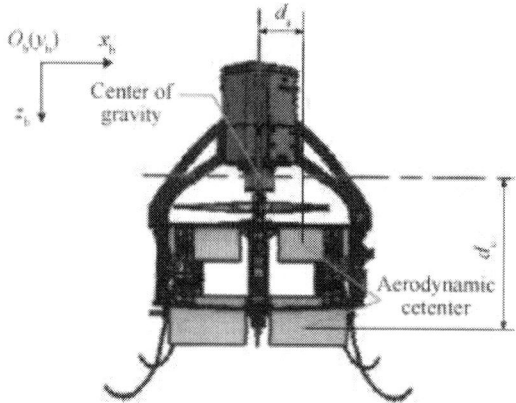

Figure 2. Components and structure diagram.

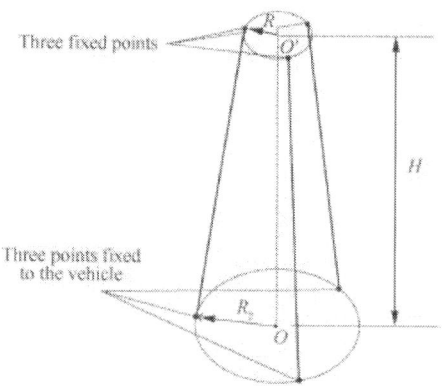

Figure 3. Tri-wire pendulum method diagram.

Where R_2 the distance from the underside suspension point to the central axis OO', R_1 the distance from the upper suspension point to the central axis OO', H the length of the suspension wire, T_0 the swing period. When the aerial vehicle takes the simple harmonic motion along OO' axis, the time for one complete cycle of any point in the aerial is defined as the swing period.

At the beginning of the experiment, the aerial vehicle is still, then it is rotated for $5°$ along the central axis OO', and the aerial vehicle is released to make a fixed-axis rotation around the central axis OO'. The rotation follows the law of simple harmonic motion. A stopwatch is used to record the time needed to accomplish 30 swing periods. The experiment is repeated for 5 times along each direction of the axes of the airframe, and all experimental data are recorded. The results in Table 3 can be obtained based on the above mentioned identification results and actual measurement.

Table 3. Experimental parameters.

Parameter	Value
Mass of the aerial vehicle m (kg)	2.81
Distance from the upper suspension point to the central axis R_1 (m)	0.12
Distance from the underside suspension point to the central axis R_2 (m)	0.18
Vertical distance of the suspension center H (m)	3.35

The experimental data for rotational inertia I_z are shown in Table 4. The values of I_x and I_y can be obtained by the same method.

Table 4. Experimental data of I_z measurement.

Time	30 swing periods
1	79.7
2	79.3
3	80.1
4	79
5	79.6
Mean value	79.7

So far, we have completed the identification of the structural parameters. The identification results of the structural parameters are shown in Table 5.

Table 5. Identification results.

Parameter	Value
Mass of the aerial vehicle m (kg)	2.81
Arm of the reaction torque flaps d_a (m)	0.08
Arm of controlling flaps d_c (m)	0.2
Rotational inertia of axis $O_b x_b I_x$ (kg·m^2)	0.0391
Rotational inertia of axis $O_b y_b I_y$ (kg·m^2)	0.0391
Rotational inertia of axis $O_b z_b I_z$ (kg·m^2)	0.0304

Identification of aerodynamic parameters

Parameter identification related to rotor dynamics

The dynamic system of the ducted fan aerial vehicle mainly contains the motor, the propeller, the electronic speed regulator, and batteries. The thrust output of the dynamic system, the torque and the slip velocity caused by the propeller are given, respectively, by

$$\begin{cases} f_r = \begin{bmatrix} 0 & 0 & -K_f n^2 \end{bmatrix}^T \\ \tau_r = \begin{bmatrix} 0 & 0 & K_t n^2 \end{bmatrix}^T \\ V_e = K_v n \end{cases} \tag{19}$$

where K_v is the slip velocity coefficient of the dynamic system.

The parameters K_f, K_t and K_v are to be identified. Based on the sources and features of these parameters, a thrust and torque testing platform was designed for the ducted fan aerial vehicle. [15] The schematic diagram is shown as Fig. 4.

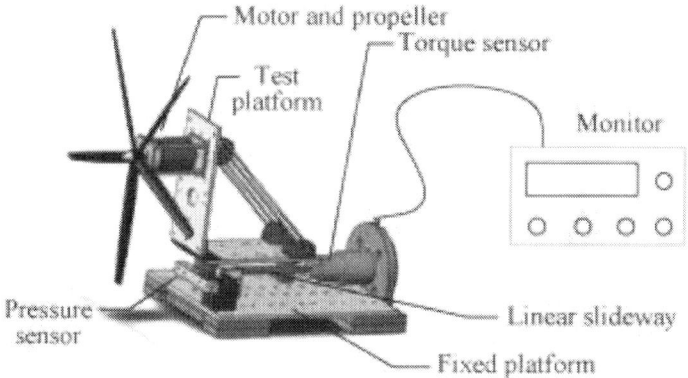

Figure 4. Multiple-sensor testing platform.

Hence, this testing system is composed of an active testing platform and a fixed platform as shown in Fig. 4. The active platform combines with the fixed one by the cylindrical linear guide rail. The measured motor and propeller are installed on the active platform. During the testing experiment, motor can make the propeller spin, and apply the pulling force and the torque to the active platform. At the same time, the pressure sensor and torque sensor can measure the pulling force and moment through the cylindrical linear guide rail. The galvanometer, the anemometer, and the tachometer are all behind the propeller, and it is convenient to test the current of the motor and measure the rotating speed and the sliding speed of the propeller timely. This testing system integrates many types of measuring sensors, including the pressure sensor, the torque sensor, the anemometer and the tachometer. The experimental apparatus for testing the dynamic system is shown in Fig. 5.

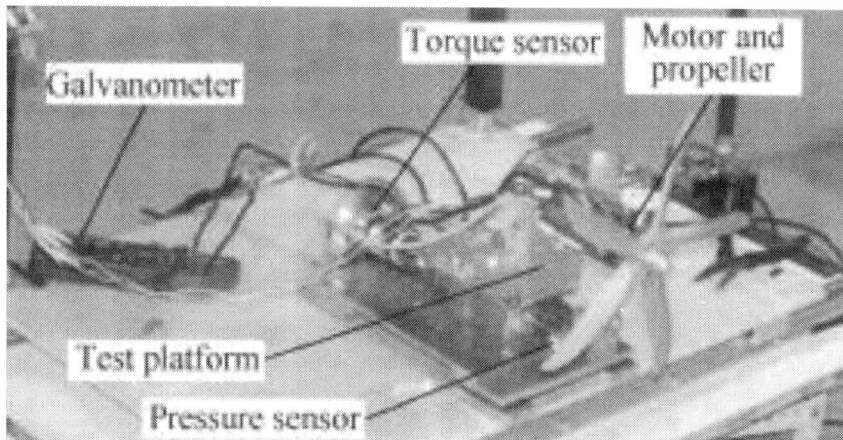

Figure 5. Testing experiment of dynamic system.

The data for the testing of the dynamic system are recorded in Table 6.

Table 6. Testing results of dynamic system.

Time (ms)	Rotating speed (r/min)	Thrust (N)	Torque (N·m)	Wind speed (m/s)
1.1	430	0.4	0.01	2.3
1.2	1438	2.1	0.025	4.7
1.3	2538	4.6	0.057	6.9
1.4	3559	8.2	0.105	9.1
1.5	4572	13.1	0.141	10.8
1.6	5642	18.3	0.215	13.4
1.7	6750	26.1	0.302	16.1
1.8	7590	33.2	0.391	18.1
1.9	8230	40.2	0.483	19.2

Data fitting is conducted for the testing results. The fitting results are as follows:

(a) Data fitting result of the throttle control and the rotating speed

$$n=1\times10^{4}(e_t-1.05)$$

where e_t is the time of high level for pulse width modulation (PWM) signal. Here, the motor is controlled by PWM signal.

(b) Data fitting result of the thrust and the rotating speed

$$F_b^z = 5.97 \times 10^{-8}n^2$$

i.e. the aerodynamic parameter is

$$K_f=5.97\times10^{-8}$$

(c) Data fitting result of the torque and the rotating speed

$$\tau_b^z = 7.6 \times 10^{-9}n^2$$

i.e. the aerodynamic parameter is

$$K_t=7.6\times10^{-9}$$

(d) Data fitting result of slip velocity and the rotating speed

$$V_e=2.17\times10^{-3}n$$

Where V_e is measured by the anemometer which is set behind the propeller, i.e. the aerodynamic parameter is

$$K_v=2.17\times10^{-3}$$

Parameter identification for drag and lift
The attitude control flaps of the ducted fan aerial vehicle are located within the flow field of the wake flow caused by the propeller in the duct. Due to the ring effect of the duct, the features of the flow field within the duct are clearly different from that of an open flow field. Therefore, the parameters for traditional airfoil vehicles, such as the lift coefficient and the drag coefficient, are not applicable in this situation, and the parameter identification needs to be re-conducted.

(1) Conditions for aerodynamic parameter identification

This study adopts the method of finite element simulation to conduct a simulation analysis on the aerodynamic features of the duct control flaps. A momentum source with the same diameter as the propeller is used as the source of thrust, and the attitude control flaps of the simulation model are set the same as that of the practical model of the aerial vehicle in order to reflect accurately the aerodynamic features of the two layers of the attitude control flaps.[16] The simulation experimental results are presented in Fig. 6.

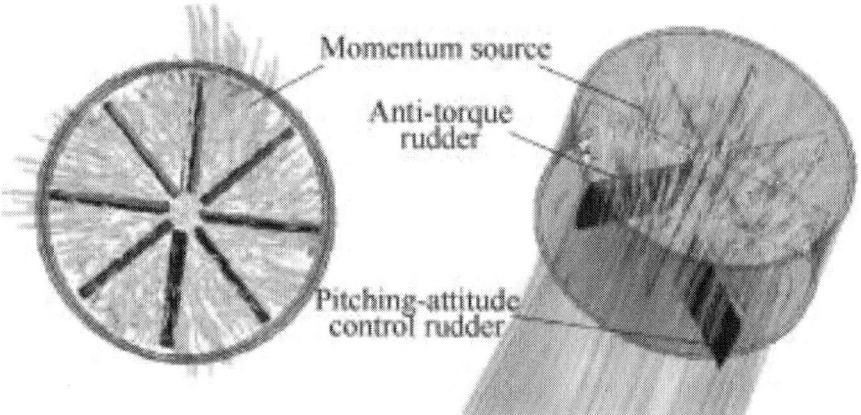

Figure 6. Finite element simulation experiment of the attitude control flaps.

The main parameters of the simulation experiment and the parameter setting methods are as follows:

a) The sizes of the reaction torque flaps and the pitching and rolling control flaps are preset in accordance with the originally designed value, $S_a = 0.0096$ m^2 and $S_o = 0.01$ m^2, respectively.

b) In case of hovering and low-speed level flight, the value of thrust can be considered to be equal to the total mass of the aerial vehicle, that is 2.81 kg, and in this situation, the wind speed in the flow field within the duct is 16.7 m/s.

c) The ideal working state of the reaction torque flap requires that when the angle of deflection equals 0°, the life can exactly overcome the spin torque of the dynamic system. The reaction torque flaps are symmetrically designed, and the force situations are all the same. Experiments will be carried out to examine the force situation of only one of the flaps.

d) The pitching and rolling control flaps are symmetric airfoils. Their lift and drag have nothing to do with the direction but with the angle of the deflection of the flaps. Experiments will be carried out to analyze the force situations of one pitching and rolling flap at the angle of the deflection at 0°, 2°, 4°, 6°, 8°, 10°.

Data of the simulation experiments are shown in Table 7.

(2) Identification results of the aerodynamic parameters of the reaction torque flaps

Table 7. Simulation experiments of attitude control flaps.

Deflection angle of the control flap	Forces of pitching and rolling control flaps (N)		Forces of reaction torque flaps (N)	
	Lift	Drag	Lift	Drag (10^{-2})
0°	0.017	0.121	0.216	0.306
2°	0.305	0.141	0.221	0.311
4°	0.583	0.161	0.219	0.308
6°	0.793	0.214	0.213	0.31
8°	1.155	0.249	0.226	0.313
10°	1.392	0.355	0.218	0.321

As indicated in paragraph (c), the ideal angle of deflection of the reaction torque flaps is 0°. In this circumstance, the lift on the control surface is expected to exactly overcome the spin torque of the dynamic system of the aerial vehicle. After carrying out testing on the dynamic system of the aerial vehicle, it can be found that the lift L_a provided equals 0.219 N, and the drag D_a equals 0.312 N. The equations of the aerodynamic lift and drag provided by the reaction torque flaps are given, respectively, by

$$\begin{cases} L_a = \dfrac{1}{2}\rho S_a V_e^2 (C_{La}\delta_a + C_{La0}) \\ D_a = \dfrac{1}{2}\rho S_a V_e^2 (C_{Da}\delta_a + C_{Da0}) \end{cases} \quad (20)$$

where the parameter values are as follows: $\rho = 1.205$ kg/m², $S_a = 0.0096$m², and $V_e = 16.7$ m/s.

Substitute all parameter values into the lift and drag calculating Eq. (20) to find that

$$\begin{cases} 0.219 = \dfrac{1}{2} \times 1.205 \times 0.0096 \times 16.7^2 C_{La0} \\[2mm] 0.312 = \dfrac{1}{2} \times 1.205 \times 0.0096 \times 16.7^2 C_{Da0} \end{cases}$$

Solve the above equation to get the lift coefficient and the drag coefficient when the angle of deflection of the reaction torque flaps is zero. The results are as follows:

$$C_{La0} = 0.1358, C_{Da0} = 0.0323$$

(3) Identification results for the aerodynamic parameters of the pitching and rolling control flaps

The airfoil profiles and sizes of the four pitching and rolling flaps are the same. During the simulation experiments, we only need to examine the force situation of one control flap. The aerodynamic lift on the surface of the control flap can be obtained by using the finite element simulation method. Data fitting for the aerodynamic lift leads to

$$L_c = 0.1432\delta_c + 0.0005 \tag{21}$$

Therefore, the expression of the aerodynamic lift of the control flap is as follows:

$$L_c = \frac{1}{2}\rho S_c V_e^2 C_{Lc}\delta_c \tag{22}$$

in which,

$$\begin{cases} \rho = 1.205 \ \text{kg/m}^2 \\ S_c = 0.01 \ \text{m}^2 \\ V_e = 16.7 \ \text{m/s} \end{cases}$$

Join the data fitting Eq. (21) with the lift Eq. (22) to get

$$\frac{1}{2}\rho S_c V_e^2 C_{Lc}\delta_c = 0.1432\delta_c$$

That is to say:

$$C_{Lc}=8.5\times10^{-2}$$

By the same procedure, coefficients related to drag can be calculated:

$$C_{Dc}=1.3\times10^{-3}$$

$$C_{Dc0}=7.47\times10^{-2}$$

So far, the identification of aerodynamic parameters has basically been completed. The identification results of the aerodynamic parameters are presented in Table 8.

Table 8. Identification results.

Parameter	Value
Thrust coefficient of the propeller K_f	5.97×10^{-8}
Torque coefficient of the propeller K_t	7.6×10^{-9}
Coefficient of the slip velocity K_v	2.7×10^{-3}
Lift coefficient of the anti-torque flap at $0°$ deflection C_{La0}	1.358×10^{-1}
Drag coefficient of the anti-torque flap at $0°$ deflection C_{Da0}	3.23×10^{-2}
Lift coefficient of the control flap C_{Lc}	8.5×10^{-2}
Drag coefficient of the control flap C_{Dc}	1.3×10^{-3}
Drag coefficient of the control flap at $0°$ deflection C_{Dc0}	7.47×10^{-2}

Dynamic model discussion
(1) Overall dynamic equation
The above parameters are substituted into the dynamics Eq. (5) to obtain the following flight dynamics model for single-blade ducted aerial vehicles:

$$
\begin{bmatrix}
-mg\sin\theta + \rho S_c V_e^2 C_{Lc}\delta_{c2} \\
mg\cos\theta\sin\phi - \rho S_c V_e^2 C_{Lc}\delta_{c1} \\
mg\cos\theta\cos\phi - K_f n^2 + 4\rho S_a V_e^2\left(C_{Da}\delta_a^2 + C_{Da0}\right) + \\
\rho S_c V_e^2\left[C_{Dc}\left(\delta_{c1}^2 + \delta_{c2}^2\right) + C_{Dc0}\right]
\end{bmatrix}
$$

$$
=
\begin{bmatrix}
m(\dot{u} + qw - rv) \\
m(\dot{v} + ru - pw) \\
m(\dot{w} + pv - qu)
\end{bmatrix}
\tag{23}
$$

$$\begin{bmatrix} \rho S_c V_e^2 C_{Lc} \delta_{c2} d_c - I_p \omega_e q \\ \rho S_c V_e^2 C_{Lc} \delta_{c1} d_c + I_p \omega_e p \\ K_t n^2 - 4\rho S_a V_e^2 (C_{La} \delta_a + C_{La0}) d_a \end{bmatrix} = \begin{bmatrix} I_x \dot{p} + qr(I_z - I_y) \\ I_y \dot{q} + rp(I_x - I_z) \\ I_z \dot{r} + pq(I_y - I_x) \end{bmatrix} \tag{24}$$

(2) Attitude control model

The paper is mainly about the attitude model of the ducted fan aerial vehicle, and gives the relevant designs of the control flaps. The dynamic equations of the pitch and roll attitude loop are as

$$\begin{cases} L = I_x \dot{p} + qr(I_z - I_y) \\ M = I_y \dot{q} + rp(I_x - I_z) \end{cases} \tag{25}$$

According to Eq. (24), the corresponding aerodynamic moments can be derived as

$$\begin{cases} L = \rho S_c V_e^2 C_{Lc} \delta_{c2} d_c - I_p \omega_e q \\ M = \rho S_c V_e^2 C_{Lc} \delta_{c1} d_c + I_p \omega_e p \end{cases} \tag{26}$$

In applying the assumption of the small-disturbance theory, Eq. (25) can be simplified:

$$\begin{cases} \Delta L = \dfrac{\partial L}{\partial n} \Delta n + \dfrac{\partial L}{\partial \delta_{c2}} \Delta \delta_{c2} + \dfrac{\partial L}{\partial q} \Delta q \\[2mm] \Delta M = \dfrac{\partial M}{\partial n} \Delta n + \dfrac{\partial M}{\partial \delta_{c1}} \Delta \delta_{c1} + \dfrac{\partial L}{\partial p} \Delta p \end{cases} \tag{27}$$

where ΔL and ΔM are changes of rolling moment and pitching moment; Δp and Δq are state variables, denoting the angular rates of the pitching and rolling of the aerial vehicle respectively, $\Delta \delta_{c1}$ and $\Delta \delta_{c2}$ are control variables, denoting the angles of deflection of the control flaps in the pitching and rolling directions; and Δn is the change of propeller speed.

Substitute the results of the identified parameters into Eq. (27), we can get

$$\begin{cases} \dfrac{\partial L}{\partial n} = \dfrac{\partial M}{\partial n} = 0 \\[2mm] \dfrac{\partial L}{\partial \delta_{c2}} = \dfrac{\partial M}{\partial \delta_{c1}} = 12.05 d_c S_c \\[2mm] \dfrac{\partial L}{\partial q} = \dfrac{\partial M}{\partial p} = 2.36 \end{cases} \tag{28}$$

Thus, the dynamic equations of the attitude control loop can be obtained as

$$\begin{cases} 2.41 S_c \Delta \delta_{c2} + 2.36 \Delta q = I_x \Delta \dot{p} \\ 2.41 S_c \Delta \delta_{c1} + 2.36 \Delta p = I_y \Delta \dot{q} \end{cases} \tag{29}$$

Based on Eq. (29), what the paper researched next is about the design of the control flaps.

COMPLETION OF THE DESIGN FOR THE CONTROL FLAPS USING THE MODEL OBTAINED BASED ON THE IDENTIFICATION

Design for control flaps

The effective lift surface area S_c and the length of the arm of the acting force d_c of the control flaps need to be confirmed. These are also the design variables of the design of the control flaps. The structural and aerodynamic parameters obtained in Section 3 are substituted into Eq. (29), the flight dynamics model of the aerial vehicle in Section 2, and then the method of small disturbance and linearization is used to obtain the following state space expression of the ducted aerial vehicle:

$$\begin{bmatrix} \Delta \dot{p} \\ \Delta \dot{q} \end{bmatrix} = \begin{bmatrix} 0 & 1.05 \\ 1.05 & 0 \end{bmatrix} \begin{bmatrix} \Delta p \\ \Delta q \end{bmatrix} + \begin{bmatrix} 0 & 308.18 d_c S_c \\ 308.18 d_c S_c & 0 \end{bmatrix} \begin{bmatrix} \Delta \delta_{c1} \\ \Delta \delta_{c2} \end{bmatrix} \tag{30}$$

where Δp and Δq are system state variables, denoting, respectively, the angular velocities of the pitching and rolling of the aerial vehicle. $\Delta \delta c_1$ and $\Delta \delta c_2$ are controlled by the system, denoting the angles of deflection of the control flaps in the pitching and rolling directions.

According to the actual flight data of the aerial vehicle, the aerial vehicle will be disturbed by the external environmental conditions, such as a crosswind. The disturbance quantity will cause the airframe of the ducted aerial vehicle to have an angular velocity of about 10 (°)/s. In this circumstance, assume that the control flaps of the aerial vehicle have reached their maximum controlling-position for controlling the vehicle. The state variables $\Delta p = \Delta q = 12$ (°)/s and the control variables $\Delta\delta_{c1} = \Delta\delta_{c2} = 12°$ are substituted into Eq. (30), which are obtained by flight tests. Then, the conditional expression of the controllability of the aerial vehicle is resolved to find that

$$\Delta\dot{p} = \Delta\dot{q} = 10.5 - 3698.16d_c S_c < 0$$

That is to say,

$$d_c S_c > 0.0028$$

A larger value for d_c indicates a closer distance to the bottom of the duct. For convenience when conducting the process and the experiment, it is best to make d_c as large as possible. Additionally, the increase of d_c will enlarge the space for the value design of aerodynamic chord of control flap. Therefore, in this paper, the value of d_c is set as 0.25 m, and thus the lift area of the control flap S_c is larger than 0.0144 m².

Simulation verification of the results of the pitching and rolling control flaps

The condition set for the simulation experiment is that, when $t = 1.5$s, the system is supplied with a certain quantity of disturbance to produce an angular velocity of $\Delta p = \Delta q = 10$ (°)/s. Meanwhile, the control flap must make sure to reach the maximum position in the opposite direction. That is, the controlled quantity $\Delta\delta_{c1} = \Delta\delta_{c2} = -12°$. The motion states of the model before and after parameter optimization are shown in Fig. 7.

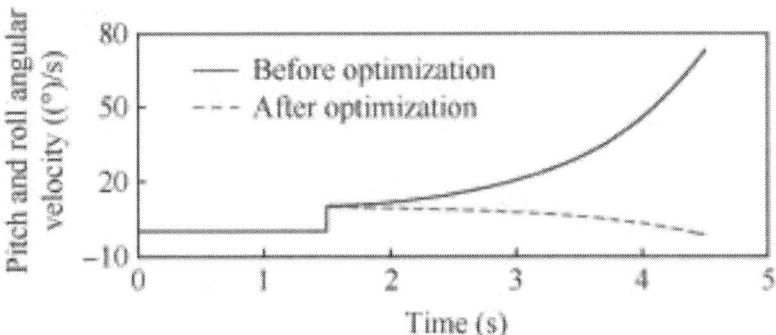

Figure 7. Simulation of restoring tendency after disturbing aerial vehicle.

According to the simulation diagram, with respect to the model of the aerial vehicle before the optimization, even when the control flap reaches the maximum steerage, it is still impossible to prevent the aerial vehicle from deviating from the equilibrium position, and as time elapses, the instability phenomenon of the aerial vehicle becomes more and more significant. After parameter optimization, when the control flap reaches the maximum steerage, the aerial vehicle shows a tendency towards a stable state, and as time elapses, such a tendency becomes more and more significant. Therefore, it improves the aerial vehicle's capacity to resist disturbance from the architectural perspective.

Flight test verification

Flight tests, shown in Fig. 8, are carried out to observe and verify the overall effect after the optimization of the attitude control units.

(a) Outdoor   (b) Indoor

Figure 8. Outdoor and indoor flight tests of the ducted fan aerial vehicle.

State parameters of the aerial vehicle are sent back using a wireless module to carry out an off-line analysis. The changes for one flight are shown in Fig. 9.

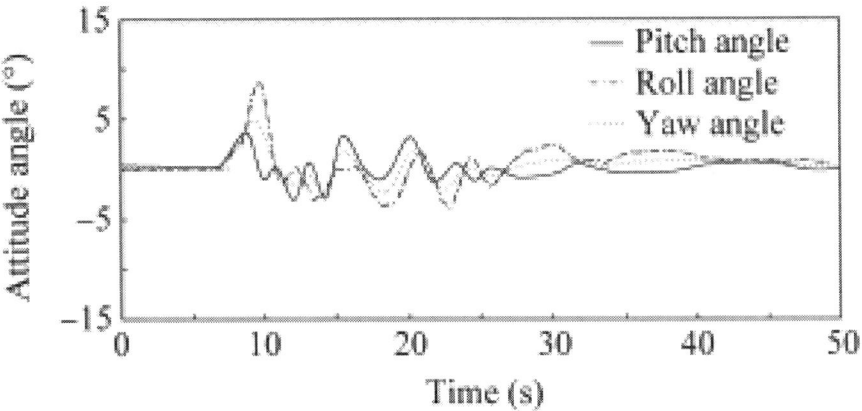

Figure 9. Flight attitude situation during the flight verification experiment.

According to Fig. 9, the aerial vehicle is able to return to the stable state gradually when it is disturbed by the external environment. It means that the design of the control structures of the aerial vehicle can significantly improve the controllability of the aerial vehicle.

CONCLUSIONS

(1) A method that dynamically integrates the structural design and the control system design of the ducted aerial vehicle is provided. The aerial vehicle architecture is carried out based on the requirements of the structural parameters using the control system of the aerial vehicle.
(2) A mathematical model with better controllability using proper structural design is obtained. Through the flight test, the accuracy of modeling and identification processes for this small aerial vehicle is verified.

ACKNOWLEDGEMENT

This study was supported by the National Natural Science Foundation of China (No.11102019).

REFERENCES

1. Cycon JP, Hunter DH, Krauss TA. Snubber assembly for a rotor assembly having ducted, coaxial counter-rotating rotors. Hartford, CT: United Technologies Corporation; United States Patent US5340279; 1994.
2. Graf WE. Effects of duct lip shaping and various control devices on the hover and forward flight performance of ducted fan UAVs [dissertation. Blacksburg, VA: Virginia Polytechnic Institute and State University; 2005.
3. Metni N, Pflimlin JM, Hamel T, Soueres P. Attitude and gyro bias estimation for a flying UAV. In: IEEE/RSJ International Conference on Intelligent Robots and Systems; 2005. p.1114–20.
4. Ohanian OJ, Karni ED, Londenberg WK, Gelhausen PA, Inman DJ. Ducted-fan and moment control via steady and synthetic jets. J Aircraft 2011;48(2):514–26.
5. Pflimlin JM, Soueres P, Hamel T. Waypoint navigation control of a VTOL UAV amidst obstacles. In: IEEE/RSJ International Conference on Intelligent Robots and Systems, 2006 Oct 9–15; Bejing, China; 2006. p. 3544–9.
6. Fantail VTOL miniature UAV [Internet]. Singapore: Singapore Technologies Aerospace (ST Aero); 2006 July 26 [cited 2011 Nov 18]. Available from: http://defense-update.com/products/f/ fantail.htm.
7. Marconi L, Naldi R, Sala A. Modeling and analysis of a reducedcomplexity ducted MAV. In: 14th Mediterranean Conference on Control and Automation 2006; 2006. p. 1–5.
8. Naldi R, Marconi L, Sala A. Modelling and control of a miniature ducted-fan in fast forward flight. In: American Control Conference, 2008 June 11–13; Seattle, WA, USA; 2008. p. 2552–7.
9. Ko A, Ohanian JO, Gelhausen P. Ducted fan UAV modeling and simulation in preliminary design; 2007. Report No.: AIAA-2007- 6375.
10. Akturk A, Camci C. Influence of tip clearance and inlet flow distortion on ducted fan performance in VTOL UAVs [Internet]. Available from: http://www.personal.psu.edu/faculty/c/x/cxc11/ papers/AHS_2010_Forum66_AA_CC.pdf.
11. Akturk A, Camci C. A computational and experimental analysis of a ducted fan used in VTOL UAV systems [Internet]. 2011 Feb 23[cited 2011 Nov 18]. Available from: http://www.personal.psu.edu/cxc11/publications.html.
12. Anderson JD. Fundamentals of aerodynamics. 4th ed. New York: McGraw-Hill; 2005.
13. Sudani N, Kanda H, Sato M, Hitoshi M, Kenichi M, Susumu T. Evaluation of NACA0012 airfoil test results in the NAL twodimensional transonic wind tunnel. 1st ed. Tokyo: National Aerospace Laboratory; 1991.

14. Song Ch, Pan J, Ye Y, Zhuang B. Measuring moment of inertia using tri-linear pendulum and its experimental error. J Mech Pract 2003;1:13–5 [Chinese].
15. Wang ZJ, Guo SJ, Li Ch. Numerical analysis of aerodynamic characteristics for the design of a small ducted fan aircraft. Proc Inst Mech Eng Part G: J Aerosp Eng 2013;227(10):1571–82.
16. Veldhuis LLM, Nebiolo S. Analysis of calculated and measured wake characteristics of a propeller-wing model; 2000. Report No.: AIAA-2000-0908.

CITATION

Zhengjie Wang, Zhijun Liu, Ningjun Fan, Meifang Guo, Flight dynamics modeling of a small ducted fan aerial vehicle based on parameter identification, Chinese Journal of Aeronautics, Volume 26, Issue 6, December 2013, Pages 1439-1448, ISSN 1000-9361, http://dx.doi.org/10.1016/j.cja.2013.10.006.

CHAPTER 3

Numerical Investigation of Flow Separation Behavior in an Over-Expanded Annular Conical Aerospike Nozzle

Miaosheng He, Lizi Qin, Yu Liu

School of Astronautics, Beihang University, Beijing 100191, China.

ABSTRACT

A three-part numerical investigation has been conducted in order to identify the flow separation behavior—the progression of the shock structure, the flow separation pattern with an increase in the nozzle pressure ratio (NPR), the prediction of the separation data on the nozzle wall, and the influence of the gas density effect on the flow separation behavior are included. The computational results reveal that the annular conical aerospike nozzle is dominated by shock/shock and shock/boundary layer interactions at all calculated NPRs, and the shock physics and associated flow separation behavior are quite complex. An abnormal flow separation behavior as well as a transition process from no flow separation at highly over-expanded conditions to a restricted shock separation and finally to a free shock separation even at the deign condition can be observed. The complex shock physics has further influence on the separation data on both the spike and cowl walls, and separation criteria suggested by literatures developed from separation data in conical or bell-type rocket nozzles fail at the prediction of flow separation behavior in the present asymmetric supersonic nozzle. Correlation of flow separation with the gas density is distinct for highly over-expanded conditions. Decreasing the gas density or reducing mass flow results in a smaller adverse pressure gradient across the separation shock or a weaker shock system, and this is strongly coupled with the flow separation behavior. The computational results agree well with the experimental data in both shock physics and static wall pressure distribution at the specific NPRs, indicating that the computational methodology here is advisable to accurately predict the flow physics.

INTRODUCTION

The flow separation in supersonic convergent–divergent nozzles is a basic fluid-dynamics phenomenon that occurs at a certain nozzle pressure ratio (NPR), resulting in the presence of shock waves and shock/boundary layer interactions inside nozzles. It has been the subject of various experimental and numerical studies in the past. Today, with the renewed interest in supersonic flights and space vehicles, the subject has become increasingly important, especially for aerospace applications for rockets, missiles, supersonic aircraft, etc. There has been a widespread desire to investigate features with shock/boundary layer interactions in highly over-expanded rocket nozzles, since these interactions are responsible for acoustic, vibrate-acoustic, thermal, and mechanical-induced loads that act on the structure. Conventional bell-type nozzles suffer from the above interactions with reduced engine performance at low-altitude highly over-expanded conditions due to fixed geometries. Based on the above background, different types of nozzle concept with altitude-adapting capabilities have been developed and tested on the ground in the past, and the aerospike nozzle is included as a strong contender for the propulsion system of reusable spacecraft.[1 and 2] In the recent years, a renewed interest in the aerospike nozzle flowfield has been generated for both rocket and aeronautic applications.[2, 3, 4, 5 and 6] However, while flow separation in over-expanded planar or bell ideal and optimized contour nozzles has been widely investigated to elucidate the phenomenon of boundary layer separation and shock interactions, aerospike nozzles have received little attention in the frame of this study. Verma[4] and Kapilavai et al.[7] presented pioneering work upon flow separation behavior in aerospike nozzles that operated at NPRs below 10, and both of the two studies indicated that the presence of flow phenomenon was associated with nozzle flow separation, and unsteady shock oscillation induced by the interaction of the shock/boundary layer seen in conventional supersonic nozzles with diverging sections could also be expected in aerospike nozzles at off-design operating NPRs. In spite of few rare studies on this subject, understanding of the flow separation behavior as well as fundamental knowledge of supersonic flow physics in the presence of shock wave propagation, shock reflection at walls, and shock/shock and shock/boundary layer interactions in such a convergent-divergent nozzle are still needed.

Studies of flow separation in supersonic nozzles dated back to the 1960s with the first work of Arens and Spiegler[8], who published the first approach to include the Mach number influence in the theoretical prediction of free shock separation. Schmucker[9]continued the research through the later years in the 1970s, and based on a simplified boundary layer integral approach, Schmucker proposed the famous purely empirical criterion for free shock separation (FSS) which is still widely used to date. From several experimental studies, performed on either full-scale[10] or subscale[11, 12 and 13] optimized nozzles, and corroborated by different numerical simulations[14, 15, 16, 17 and 18], the presence of two distinct flow separation patterns, namely FSS and restricted shock separation (RSS), is demonstrated. A transition in the separation pattern from FSS to RSS and vice-versa might occur, which was firstly observed in the early 1970s by Nave and Coffey.[10] At the initial state of start-up or when a supersonic nozzle operates at low NPRs, the flow mostly resides in an FSS state[19], as shown in Fig. 1.[20] A single incipient separation of the flow along the interior surface of the nozzle is triggered by an adverse pressure gradient between the regions of isentropic expansion and subsonic entrainment. The shock that originates from the incipient separation line interacts with a reflected shock; this shock emanates from the triple-point which is the location where the Mach disk, internal and reflected shocks coincide. In the FSS state, a separation region forms which encompasses a series of compression/expansion waves, and this separated flow fails to reattach back to the wall at low NPRs due to the lack of outward radial momentum as a free supersonic annular jet. A recirculating subsonic region forms between the separated free annular jet and the nozzle wall, which entrains ambient air along the nozzle wall and adapts the static wall pressure to the ambient condition.

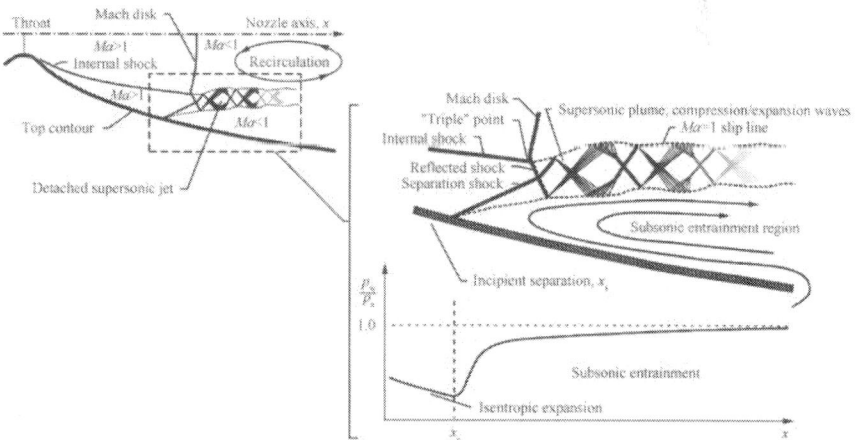

Figure 1. Illustration of the internal shock structure in a thrust optimized parabolic nozzle during an FSS state.[20]

The RSS state refers to the canonical shock/boundary layer interaction which is present in many high-speed devices. For example, RSS is known to occur in thrust optimized parabola nozzles of engines like Vulcain, space shuttle main engine (SSME), or J-2S during the start-up process. RSS is characterized by a small separation region or bubble which exists immediately downstream from the shock wave, in which the mean flow circulates and separates or tilts away from the wall before the flow reattaches and continues down the length of the nozzle as an attached boundary layer. Depending on the nozzle contour and expansion ratio, more than one bubble may be present, as shown in Fig. 2.[20] Upon further increases in the NPR, the RSS flow regime will translate downstream and the enclosed separation bubble opens up to the ambient environment when passing over the nozzle lip. When the last annular separation bubble downstream from the incipient separation shock opens up, the flow structure then switches to an FSS state due to the presence of a single separation shock with an associated separated flow region downstream.

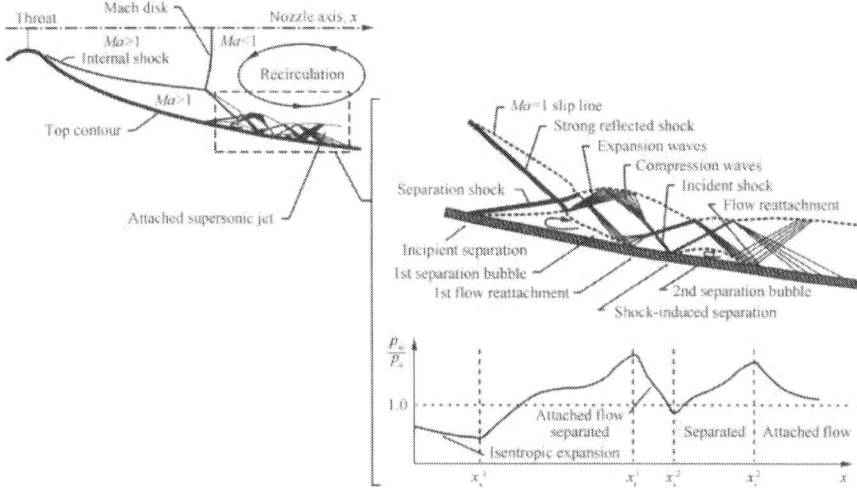

Figure 2. Illustration of the internal shock structure in a thrust optimized parabolic nozzle during an RSS state.[20]

In the past, many researchers from different groups in Europe[21 and 22], USA[10], and Japan[23] have distinguished these flow separation behaviors in rocket nozzles and have identified a transition phenomenon between the two separation regimes during the start-up or shut-down process. In a typical rocket engine, the combustion chamber pressure rises from the ambient pressure to the steady-state operating condition[14, 17 and 24], and flow separation occurs when the chamber pressure is relatively low, so as to yield a static wall pressure much lower than the ambient one in some locations of the nozzle divergent. During the start-up process, the nozzle flow is popularly dominated by the FSS structure, and then when the combustion chamber pressure rises over a certain critical value, the FSS regime is replaced by the RSS regime. The identification of this transition is important for nozzles applications in rocket engines or supersonic aircraft, as it is directly attributed to the largest lateral loads seen during their operations.[16 and 25] Previous studies on this subject have also shown that these flow features in the presence of FSS/RSS transition, unsteady shock motion behavior, and side loads are strongly dependent on the nozzle geometry, the strength of the shock, and the over-expanded ratio. A recent paper by Ostlund and Muhammad-Klingmann[26] reviewed several conditions and regimes which led to the most severe side-load[10], and commented on the difficulty in accurately modeling these phenomena.

In the case of aerospike nozzles, although they have been extensively studied experimentally and numerically since the 1950s[27, 28, 29, 30 and 31], particular attention has been mostly paid to the so-called altitude-compensating capacity.[15, 16 and 28] A considerable body of theoretical and experimental data indicates that this altitude-compensating capacity enables an aerospike nozzle to eliminate over-expansion losses altitudes below the design point, and to provide a gain in performance relative to a conventional convergent-divergent supersonic nozzle.[5] To get the highest benefit with this nozzle concept, the design pressure ratio or the design geometrical area ratio is always chosen as high as possible. Therefore, for rocket or supersonic applications, the lower operating NPRs during the start-up process can also result in the shock structure to reside within the internal nozzle exit or on the spike surface, and the presence of flow phenomenon associated with nozzle flow separation seen in supersonic nozzles with diverging sections can be expected in aerospike nozzles.

Verma et al.4[, 32 and 33] presented a series of experimental studies of shock physics and nozzle performance in two classic aerospike nozzle concepts, and the conical annular and linear with low-angle plug configurations, truncated and full length with or without freestream effects were also considered. For an over-expanded 15° annular conical aerospike nozzle[4], it was observed that the over-expansion shock from the internal nozzle, the over-expansion shock on the spike surface, and the expansion fan from the cowl lip of the internal nozzle dominated the overall flowfield development. Increasing the NPR changes the angles of these shocks as the internal nozzle operates from the over-expanded to the under-expanded condition. This produces three different types of flow separation conditions on the spike. The paper gave an idea of the flowfield once the shock moved out of the internal nozzle, while the flow features when the shock structure resided within the internal nozzle section at lower NPRs were not available. However, complex shock physics and flow features can be expected in this spectrum of NPRs as indicated by present numerical studies. Kapilavai et al.[7] dedicated his study to the aerodynamic characteristics of a so-called shrouded plug nozzle at various operating conditions, and the nozzle was also observed to be dominated by shock/shock and shock/boundary layer interactions at all off-design NPRs. Depending on shock interaction with the boundary layer on the plug wall, the nozzle exhibited both fully separated and reattached boundary layer regimes. However, it should be noted that the shrouded plug nozzle whose shroud or cowl extends over a considerable portion offers both aerodynamic and structural characteristics with respective to a more conventional aerospike nozzle. These results and comments indicate that a coupled research effort between experiment and numerical simulation is needed to fully understand the shock physics in aerospike nozzles for the complete spectrum of nozzle pressure ratios.

The current work attempts to reproduce and to expand the experimental studies of Verma.[4] The aim of the paper is to numerically study in detail the flow separation behavior at sea-level as well as imposed high-altitude simulation conditions without freestream flow, for the purpose of providing an insightful understanding of the shock physics and characteristics of shock/shock and shock/boundary layer interactions at various operating conditions, from highly over-expanded conditions to the designed point of the nozzle, in particular, the shock-induced flow separation behavior and its evolution process with the change of NPR. A comparison of the separation data against a series of separation criterions form literatures is conducted to address their applicability for the prediction of flow separation behavior in such an asymmetric supersonic nozzle. Additionally, a comparison of flow separation behavior at sea-level conditions against the imposed high-altitude simulation conditions is to demonstrate the gas density effect on afore-mentioned flow characteristics in present aerospike nozzles; in particular, its effect on the flow separation behavior in highly over-expanded conditions has been studied when more complex shock/boundary layer interactions can be expected.

NUMERICAL ANALYSIS

Experimental details and modeling

The experimental work performed on the present aerospike nozzle was carried out in a 0.5-m base flow facility, a special purpose blow-down-type tunnel, detailed in Ref.[4]. The salient features relevant to this study are shown in Fig. 3, in which the 15° half-angle θannular conical aerospike nozzle model with a design Mach number of 2.0 was mounted on the central cylindrical inner body. The afterbody contour was designed based on the recommendations given in Ref. [34]. To obtain the steady pressure distribution on the spike, up to 12 pressure points with a pitch of 4.0 mm were installed at axial locations along a single line on the spike surface. The nozzle exit radius r_e is 25 mm, the annular gap at the throat section h_t is 9 mm, and the length of the spike or plug L is 59.71 mm. The aerospike nozzle area ratio is defined as the ratio of the area at the spike end A_e to the annular throat area A_t. The length of the cowl, measured as the distance from the throat section to the cowl lip, l is fixed as 9.0 mm, resulting in an area ratio of the inner nozzleε_i = 1.19.

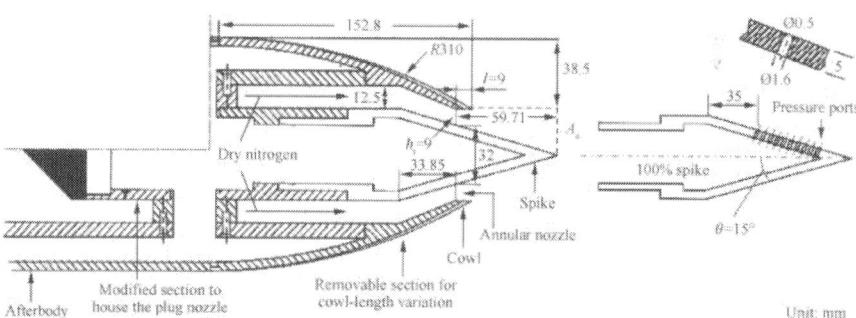

Figure 3. Schematic of the experimental annular conical aerospike nozzle model.[4]

In the present case, a modified experimental annular conical aerospike nozzle model is used, resulting in an annular axisymmetric nozzle configuration, as seen in Fig. 3, where the flow phenomena of primary interests takes place. The modified section to house the plug nozzle is not included in the axisymmetric computations; this strut is unlikely to have major effect on the nozzle flowfield as it is located in the upstream subsonic convergent section of the nozzle. A straight annular tube similar to the one in the experiments is utilized in the present numerical model to provide an area-constant expansion region until the subsonic streams reach the beginning of the nozzle convergence.

Flow conditions

The computational flow conditions reproduce and extend on the experimental conditions, that is, NPR covers a wide range of 1.60–9.87, corresponding to the unchoked subsonic condition to a slightly under-expanded condition. In the experiments, NPRs were conducted as

$2.10 \leqslant \text{NPR} \leqslant 5.75$ for the full-length spike analysis. As indicated in Section2.1, the presence of the cowl results in an additional diverging section of the shock interacting with the boundary layer on the upstream spike and cowl surface at low NPRs, where the main flow phenomena were unable to be captured by the schlieren images in the experiments. The present NPRs design will provide a complete flow expansion spectrum for such a supersonic nozzle configuration.

Three environment back pressure, p_b, or ambient pressure, p_a, conditions are designed to conduct the gas density effect analysis, as shown in Table 1. In Table 1, p_{ON} is the total pressure at the nozzle inlet. Comparisons are made between a sea-level case and two imposed high-altitude simulation cases with back pressure values of 50% and 25% of the sea-level ambient pressure condition, respectively. The NPRs conducted are identical for all three cases. The objective of the simulation is to assess the impact of the low ambient pressure environment, resulting in low nozzle total pressure and then low gas density for the NPR mimic on the shock structures and flow separation characteristics. In the present study, the Reynolds number is based on the hydraulic diameter defined as the incipient separation location ranged from 1.59×10^6 to 34.93×10^6, corresponding to a fully turbulent boundary layer.

Table 1. Condition cases for gas density effect analysis.

Ambient pressure conditions	Case No.	NPR	p_{ON} (kPa)	p_a (Pa)	Mass flow rate(kg/s)
Sea-level atmosphere	01-Jan	1.60–9.87	162.12–1000.09	101325	0.4236–2.6129
50% of sea-level ambient pressure	01-Feb	1.60–9.87	81.06–500.04	50662.5	0.2118–1.3065
25% of sea-level ambient pressure	01-Apr	1.60–9.87	40.53–250.02	25331.3	0.1059–0.6521

Numerical procedure

The numerical study has been conducted using a finite-volume unsteady Reynolds-averaged Navier–Stokes (RANS) solver. The two-equation shear stress transport (SST) model of Menter et al.[35] is used here to describe the turbulence. Previous successful efforts to compute the internal nozzle separated flow using RANS-type models by Hunter[36], Xiao et al.[37, 38 and 39], and Carlson[40] indicate that the SST model provides a reasonable prediction of flow separation just downstream from the shock caused by the shock/boundary layer interaction inside the nozzle, and therefore the best capture of the shock location and pressure distribution against the experiment. In present study, all computations are made using the axisymmetric assumption with a primary motive of a fast and efficient means of obtaining insight into the relevant shock structure and flow separation behavior at various operating conditions. Later, when the computational results are compared with the experimental data, it will be shown that the axisymmetric assumption is accurate for the prediction of salient shock physics as well as static wall pressure.

Fig. 4 shows the nozzle numerical model, including the detailed grid distribution in the nozzle region where the flow phenomena of primary interest take place. The computational domain includes the domain inside the nozzle and an ambient region around the outer surface. The domain extends $15D$ (where D is nozzle exit diameter) upstream from the throat, more than $40D$ downstream, and $10D$ in the direction perpendicular to the axis. In terms of grid structure, multi-block structured grids have been used in this calculation and are shown in Fig. 4(b). Only every second grid line in each direction is displayed for clarity. For the nozzle region, grid density is higher in the divergent part of the nozzle to improve the resolution for capturing shocks. In addition, the grid is clustered along the cowl and spike walls to resolve the boundary layers, for a Reynolds number based on an order of 10^6, and the first grid point from either the spike or the cowl wall gives a $y^+ < 1$. A detailed discussion of the grid convergence study is reported in the next section.

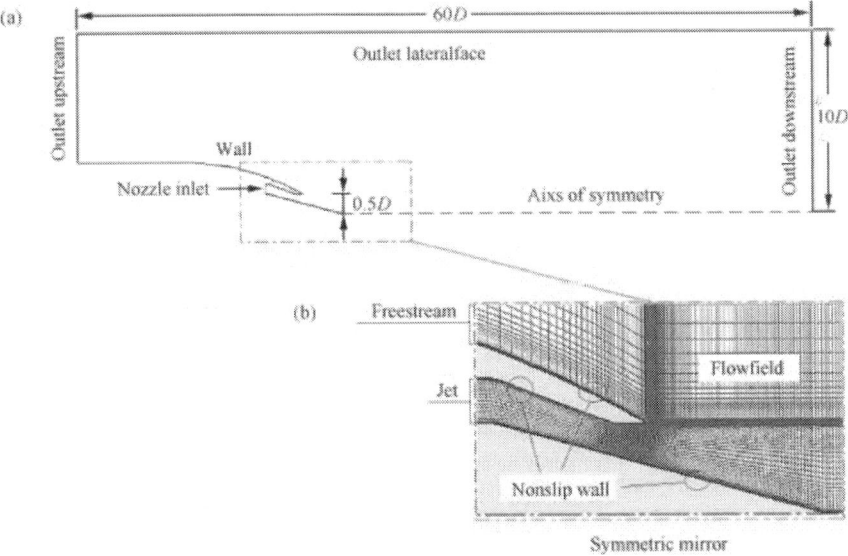

Figure 4. Illustrations of the nozzle numerical model: (a) overall computational domain, both inside and outside the nozzle, and (b) a close-up view of mesh topology in vicinity of the nozzle region.

Fixed total conditions have been employed for the outerface boundary, outlet upstream, outlet lateralface, and outlet downstream, the total pressure is set to the ambient pressure, and the boundary condition is designed in such a way that the supersonic plume may go out of the boundary, while the ambient air may come into the boundary due to the possible plume entrainment effect. The total pressure and the total temperature at the nozzle inlet are specified to be $p_{ON} = NPR \times p_a$ and $T_{ON} = T_a$, respectively. For all cases, the stagnation and the ambient temperature, T_{ON} and T_a, in the computations are fixed as 300 K, and pure gaseous nitrogen is simulated as the nozzle jet while a perfect gas with the properties of air is treated as the ambient fluid. A no-slip, adiabatic condition is specified for the solid walls, and a symmetric condition is imposed on the nozzle axisline. In addition, the upstream turbulence intensity is prescribed at the nozzle inlet boundary based on the located hydraulic diameter.

Preconditioning is employed for accelerating the calculation at low subsonic NPRs. However, the computed shock structures show the flow to assume a steady state in contrast to the experiment, in which a 1500-Hz peak in frequency can be seen in the region of separation.[4] This is particularly true when RANS-type time-averaged models are utilized to compute an internal nozzle flow with random unsteadiness or distinct acoustic tones. For the present study, all the computations are conducted with an RANS-type modeling method in the primary interest of salient shock structure and flow separation characteristics development as the NPR is increased from low to high values in such an asymmetric supersonic nozzle.

Grid convergence

In order to establish the fidelity of the numerical database, we have conducted a grid convergence study on the nozzle, and a series of three computational meshes has been run. As it is important to accurately capture the salient shock structures and their interaction with the boundary layer which drives the flow separation behavior. We begin the discussions by comparing the computational flow patterns and the static wall pressure profiles at specific NPRs with the data taken from experiments at similar NPRs for the grid convergence study. In the experiments, when increasing the NPR values, the internal nozzle operates from the over-expanded to the under-expanded condition, three types of flow separation patterns can be observed at NPR = 2.10, 2.57, 3.82, respectively.[4] The comparisons in this section are all for the above three NPR conditions. Schlieren images have been used in the experiments to help in identifying the first gradients of density, which can exhibit the shock patterns in the flowfield. To make a direct comparison of the computational flowfield with the experimental results, the computational results also use the numerical schlieren pictures contoured by the plots of the absolute values of the first gradients of density at the grid nodes.

The shock-induced flow separation patterns for various mesh resolutions are displayed in Fig. 5, in which the numerical schlieren pictures listed from the first to the third row show the results based on the coarse (Mesh A), medium (Mesh B), and fine grid (Mesh C) cases, respectively. For all the three meshes, the three types of flow separation patterns at specific NPRs are well reproduced, and the numerical results are in very good agreement since the shock structures are very close to each other.

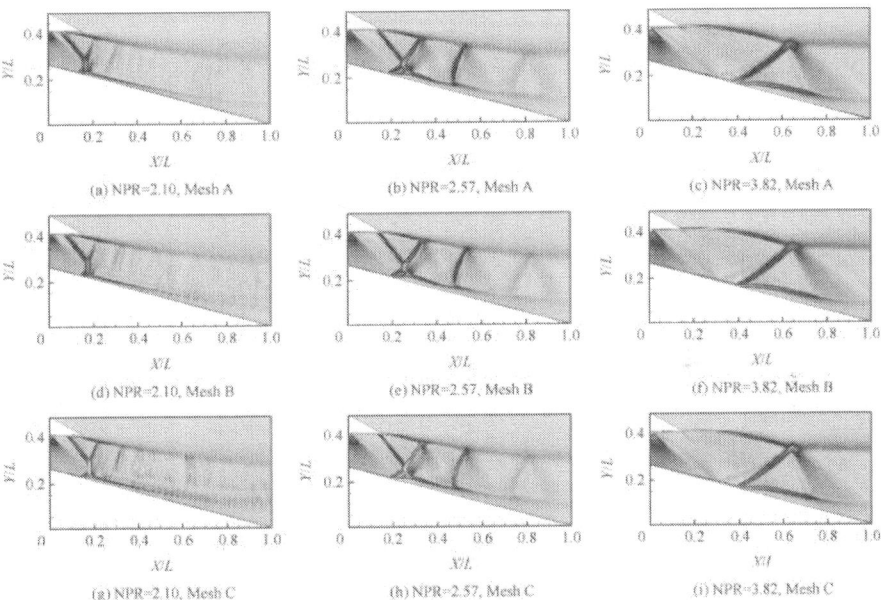

Figure 5. Comparison of shock-induced flow separation patterns for various mesh resolutions.

The pressure distributions along the nozzle spike and the cowl wall have also been compared for the three meshes. Fig. 6 shows the static wall pressure p_w profiles normalized by the ambient pressure, p_a, at the three specific NPRs of 2.10, 2.57, and 3.82, respectively. Again, the plots are really similar for the three meshes, but a slight shift is observed between the coarse grid and the two finer grids. In addition, it is shown that the results for the medium and fine meshes collapse almost perfectly. It is noteworthy that the shock structures and location as well as the separation point are well predicted whatever mesh is considered. As a consequence, the so-called fine grid mesh is used in the present study, as the shock pattern captured is more elaborate as the grid is refined in the nozzle flow region, especially in highly over-expanded conditions when more complex shock/boundary layer interaction can be expected.

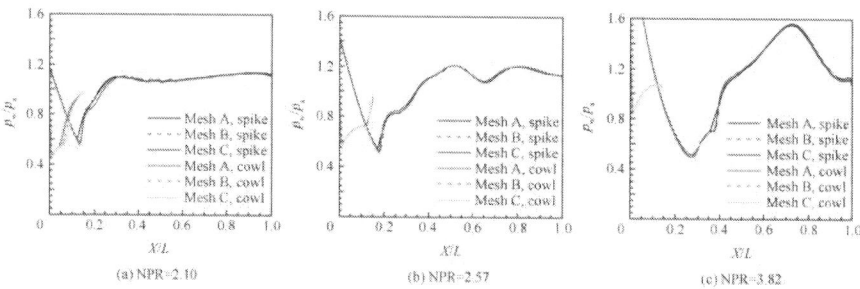

Figure 6. Static wall pressure profiles of the spike and the cowl for various mesh resolutions.

RESULTS AND DISCUSSION

Comparison with experiments and validation

Computational results for the three specific NPRs of 2.10, 2.57, and 3.82 have been compared to the data taken from the experiments for the same NPR conditions. The comparisons of flow separation patterns and static wall pressure profiles are shown inFig. 7. In Fig. 7, Plug-Num and Plug-Exp represent numerical calculation and experiment of plug. The photographs on the second row show the experimental schlieren images, while the ones on the first row show the computational results at the same NPRs. For the three NPRs, three different types of flow separation patterns as well as basic flow characteristics that can be observed are the external jet boundary developing from the cowl lip, over-expansion shocks formed on both the cowl and spike surfaces, and the interaction between them. The computational schlieren clearly replicates these salient flow features. However, an RSS characteristic, separated wake being reattached because of the Coanda effect[41], is also captured in the computation at NPR = 2.10 while an FSS condition was reported by Verma.[4] The numerical schlieren shows the presence of wake reattachment on the spike surface which cannot be distinguished in the experimental schlieren image, and a detailed analysis on the shock physics can be found in a later section. In addition, the experiments used both vertical and horizontal configurations of the knife edge in the schlieren setup. The vertical knife edge helped to identify horizontal gradients, whereas the horizontal knife edge helped to identify the vertical gradients. Note that the upper and lower halves of the experimental flow schlieren exhibit an axisymmetric nature of the shock pattern which indicates that the axisymmetric computational methodology here is advisable to accurately predict the flow physics.

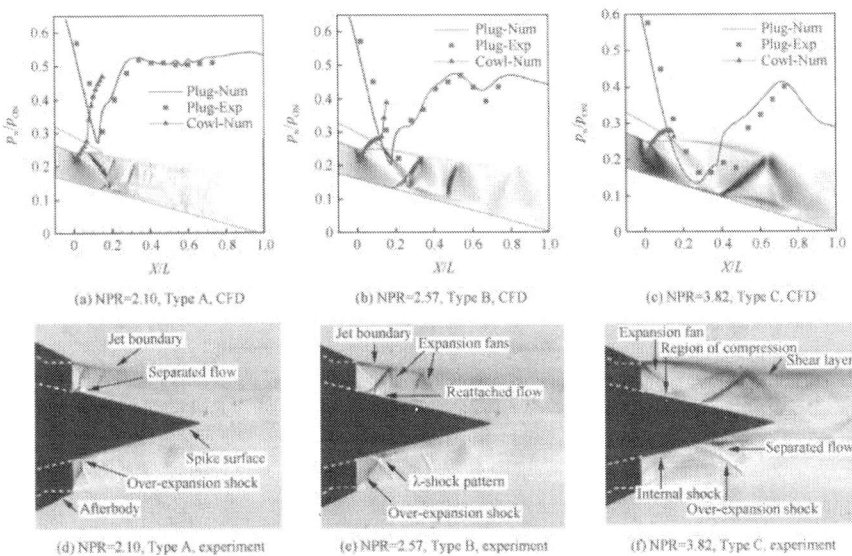

Figure 7. Comparison of static wall pressure distributions and shock-induced flow separation physics between numerical calculations and wind tunnel experiments with no freestream by Verma,[4] Case 1-1, p_a = 101325 Pa.

In terms of static wall pressure distribution, as shown in numerical schlieren images on the top row, the static wall pressure distributions along the spike surface are compared. All three plots show good agreement with the experimental results in both shock location and pressure distribution. However, the shock location is under-predicted for all the three NPRs and the miss-prediction expands as the NPR changes from low to high values. This discrepancy may contribute to miss-mimicking the experimental condition. In the experiments, the ambient pressure p_a was measured on the afterbody, 15-mm upstream from the cowl lip, by varying the jet stagnation pressure p_{oj} to vary the NPR, and the local ambient pressure p_a decreased with each increase in p_{oj} because of the jet entrainment effect. Thus, the NPR defined as the ratio of the upstream total pressure to the ambient pressure, is therefore different for the experiments and the present numerical simulations. Overall, the computational results agree well with the experimental data in both shock physics and static wall pressure distribution, and the characteristics described by both the simulations and the experiments are indicative of the flow physics that are observed at the three NPRs.

Flow separation and shock structure progression in the nozzle

To understand the details of the flow separation behavior as well as the shock structure motion in the nozzle, computations when the NPR is increased from low to high values are conducted and the numerical schlieren images of shock structures constructed from computational results are summarized in Fig. 8. In this section, the numerical schlieren pictures are also contoured by the plots of the absolute values of the first gradients of density at the grid nodes, and additionally, static wall pressure distribution plots along the cowl and spike surfaces are superimposed on the schlieren pictures in order to help understand the flow physics.

We begin the discussion with NPRs ranging from 1.60 to 1.90 when the shock structure resides within the cowl, but was not captured in the experiments.[4] A prominent feature that can be found in the computational schlieren is that a fully separated shear layer emerges from the throat of the cowl-side, and the shear layer as well as the spike surface forms an aerodynamic diverging passage downstream from the throat region. This phenomenon is absent in an over-expanded conventional nor the similar shrouded plug nozzle reported by Kapilavai et al.[7] Fig. 9 gives a line diagram of the nozzle throat design, in which the aerospike nozzle considered presently has a sharp change in the slope at the throat on the cowl-side, and a turning angle up to 20.37° makes the nozzle inlet total pressure here not high enough to expand the jet to form a wall-bounded flow on the cowl surface by Prandtl–Mayer expansion. As the NPR increases, the separated shear layer moves toward the cowl surface and the flow in the above aerodynamic diverging passage exhibits a multitude of shock structures. Detailed flow physics is seen more clearly in Fig. 10.

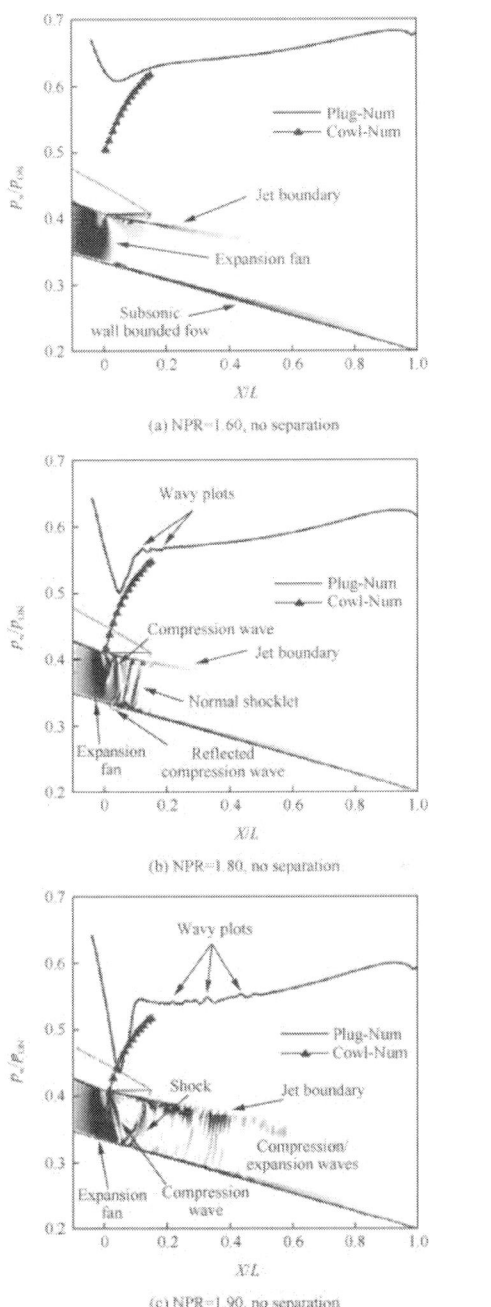

Figure 8. Shock structure motion as the NPR is increased from 1.60 to 1.90 at the sea-level atmospheric conditions, Case 1-1, p_a = 101325 Pa.

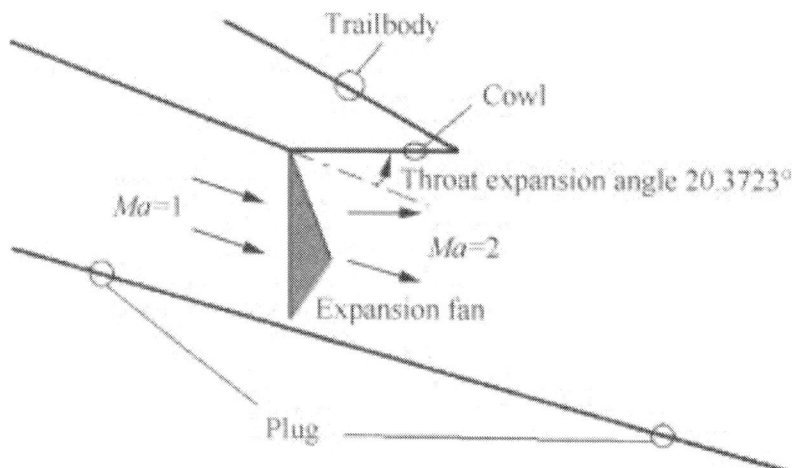

Figure 9. Line diagram of the annular conical aerospike nozzle throat configuration.

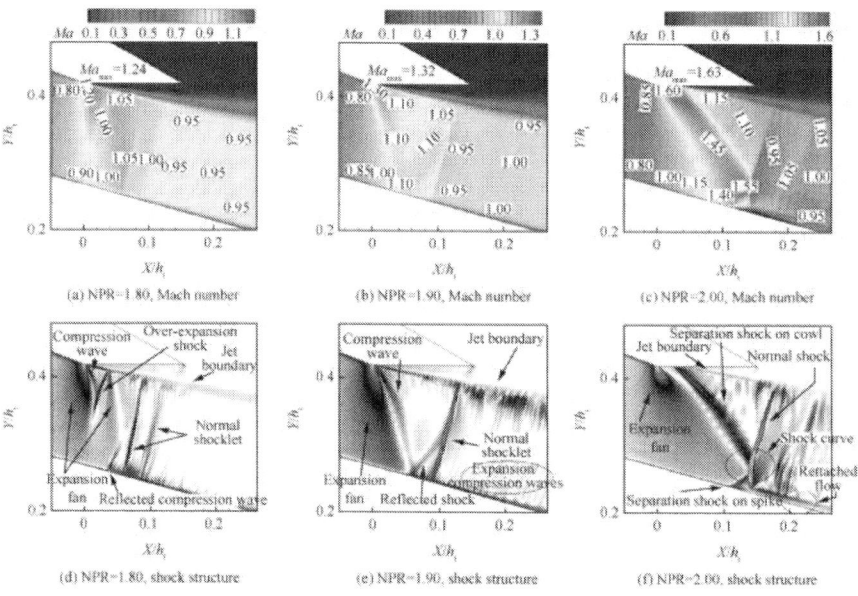

Figure 10. Close-up view of flow characteristics near the throat region at low NPR conditions for Case 1-1,p_a = 101325 Pa. Pictures are contoured by the data constructed from the computational results with the Mach number for the upper row while the absolute values of the first gradients of density at the grid nodes are on the bottom.

For the NPR = 1.60 shown in Fig. 8 (a), note that the static wall pressure at the throat region is over 0.528 times of the nozzle inlet total pressure, and the flow near the throat region is still unchoked, so no shock structure is captured in the flowfield at this NPR condition. When the flow negotiates the throat region, it goes back to a subsonic wall-bounded flow condition on the spike surface. As the NPR increases to 1.80, the throat region starts to choke as shown in Fig. 8(b) and Fig. 10(a), respectively. The shock exhibits an apparent asymmetric structure like a normal shocklet structure on the spike surface, which is similar to what can be discerned in a transonic diffuser at low NPRs[42]while a serial of expansion and compression waves or shocks appears on the cowl-side flow region. As shown in Fig. 10(a), the flow is compressed immediately by the separated shear layer just after an expansion to Mach number 1.24 through an expansion fan at the sharp corner, and then again expands to an over-expansion shock downstream. The expansion fan that emerges from the over-expansion shock foot over-expands the jet finally to a normal shocklet in the diverging section. As the NPR increases to a slightly higher value of 1.90 as shown in Fig. 8(c) and Fig. 10(b), the normal shocklets coalesce in a single normal shock downstream and occupy the entire cross section. However, the shock is still not strong enough to induce the spike wall boundary separation. In this NPR case, the compression waves induced by the separated shear layer now coalesce in an oblique compression shock which travels across the entire stream without the interception of the over-expansion shock observed in the case of NPR = 1.80. This oblique compression shock is incident on the spike surface and the reflected shock from the spike surface hits the normal shock downstream. Another feature that can be noticed is the presence of weak expansion and compression waves in the aft region of the normal shock, which seem to be a manifestation of the natural convergent–divergent passage formed downstream from the normal shock, and evidence can also be found from the wavy plots of the static wall pressure distribution along the spike surface.

The separated shear layer finally reattaches on the cowl surface when the NPR is about 1.96, and the flow undergoes shock-induced flow separation on both the cowl and spike surfaces when the NPR increases to 2.00, as shown in Fig. 11(a). A close-up view of the computational result can be found in Fig. 10(c). At this point, the shock structure consists of oblique shocks starting from both the spike and cowl surfaces, and these two oblique shocks occupy the downstream normal shock where their interaction makes the normal shock of the spike-side flow region to curve. The flow along the spike surface separates and reattaches in a short distance forming a small recirculation bubble while the flow is fully separated over the entire length of the cowl just aft of the oblique shock.

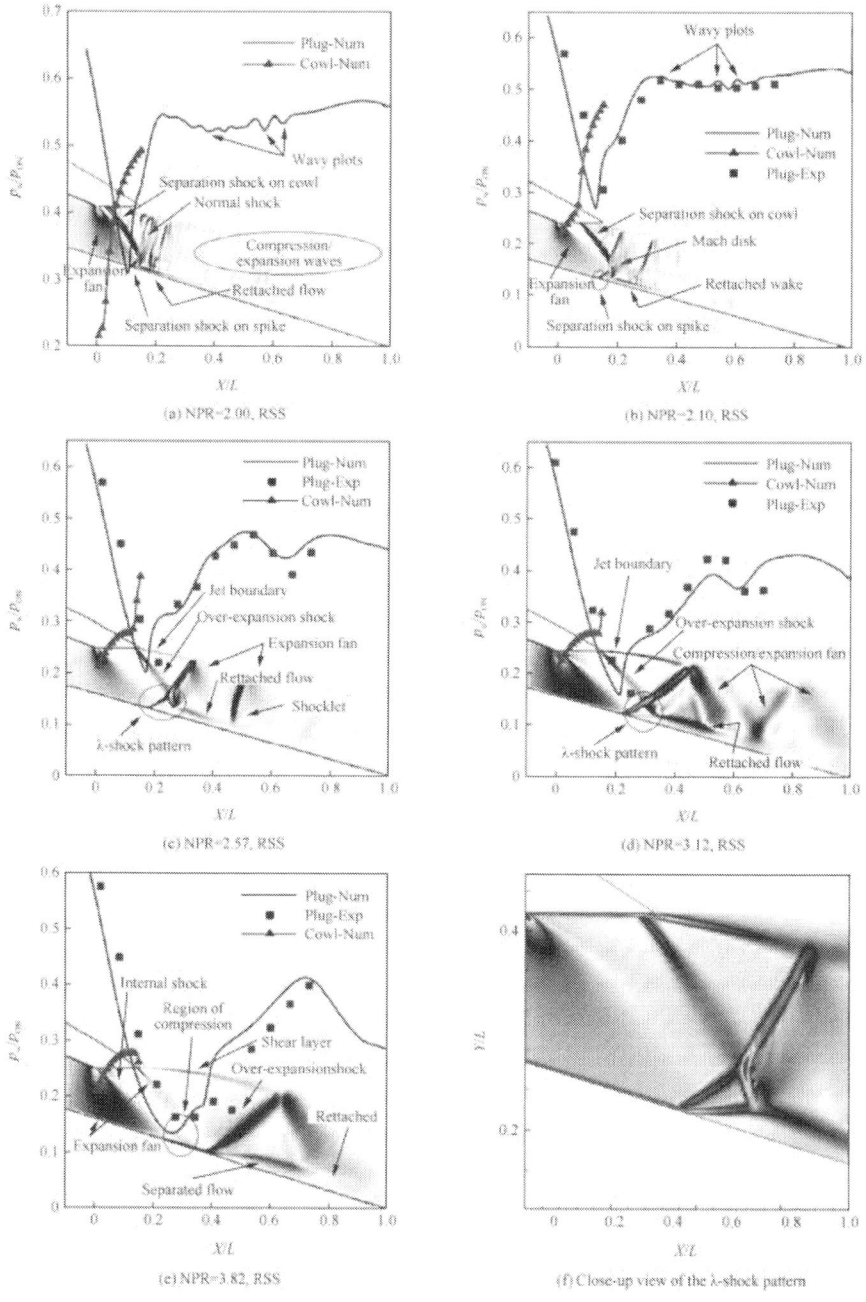

Figure 11. Shock structure motion as the NPR is increased from 2.00 to 3.82 and a close-up view of the λ-shock pattern for an NPR of 2.57 at the sea-level atmospheric conditions, Case 1-1, $p_a = 101325$ Pa.

As the NPR increases from 2.00 to a slightly higher value, the shock structure in the mean flow starts to move out of the cowl which can be captured by the experimental schlieren images. Increasing the NPR from 2.10 to 3.82 changes the angles of the oblique shocks on both the cowl and spike surfaces and their interaction at the internal nozzle operates from the over-expanded to the under-expanded condition. As indicated by Verma[4], this produces different shock structures as well as shock-induced flow separation behaviors on both the cowl and spike surfaces, which can be classified into three types. Firstly, in the low NPR = 2.10 case, the numerical schlieren shows a shock/boundary layer interaction similar to that seen in planar supersonic nozzles.[12 and 43] The shock structure exhibits a Mach reflection,[44] and oblique shocks starting from both the spike and cowl surfaces anchor a normal Mach stem in the mean flow. The flow along the spike surface separates and reattaches again in a short distance forming a small recirculation bubble, and evidence can be found in figures illustrated in a later section, which are contoured by the static wall pressure distribution and axial value of wall shear stress plots along the spike surface, thus exhibiting a restricted shock flow separation, RSS, condition. However, Verma has reported an FSS characteristic for this NPR case in Ref.[4], and this discrepancy may be produced by the undistinguishable wake reattachment in the experimental schlieren image. At this point, the flow separation behavior is not a classical RSS condition, while the mechanism of the wake reattachment is more likely the Coanda effect.[45 and 46] The asymmetric flow passage leads to a radial momentum toward the spike-side flow region and the separated shear layer is very close to the spike surface, so the lower pressure in the separation zone works like an extra suction force, in addition to that of the original Coanda effect on the supersonic jet side, ensuring the separated wake to adhere to the wall.

At an intermediate value between 2.57 and 3.12, a second shock structure can be discerned. The length of the normal portion of the normal shock decreases with the increase of the NPR, because the strength of the oblique shock on the cowl surface weakens while the interaction between the oblique shock on the spike surface and the normal shock forms a lambda shock pattern on the spike. Detailed flow physics of such a λ-shock pattern for an NPR of 2.57 is seen more clearly in Fig. 11(f). The flow then reattaches which increases the local static wall pressure above ambient, as shown inFig. 11(c)–(d). For NPR = 3.82, a third flow condition is produced when the internal nozzle starts to operate in a little under-expanded condition shown in Fig. 11(e). The expansion fan emerges from the cowl lip and over-expands the flow along the spike surface, resulting in the forming of an oblique over-expansion shock, which hits the free shear layer from the cowl lip and induces an expansion fan at the intersection point. The shock pattern on the spike surface that encloses a recirculation bubble indicates that the nozzle is still in the RSS regime. Another feature that can be discerned from the schlieren is that the internal shock originating from the sharp corner of the throat impinges on the spike surface upstream from the separation shock, forming a region of compression that induces a pressure bump shown in Fig. 11(e).

At higher NPRs, the expansion fan impinges further downstream causing the over-expansion shock as well as the flow reattachment point to move downstream. At NPR = 5.22, the over-expansion shock starts to induce a free shock separation condition on the spike, as shown in the axial wall shear stress plots in later section. The length of the separation bubble just covers the entire portion of the spike downstream from the incident separation point; detailed analysis on this flow condition is followed in a later section.

For NPRs above 5.22 through 8.40, the free shock separation is the mode of separation on the spike surface. Increasing the NPR to 5.75 continues to weaken the separation shock on the spike while strengthening the internal shock as well as the expansion fan from the cowl lip, and the region of compression induced by the impingement of the internal shock now produces a distinct pressure bump, as shown in Fig. 12(a). Additionally, an important phenomenon in these NPR cases that should not be neglected is the motion of the separated shear layer away from the spike surface. This behavior indicates a higher pressure in the separation zone to push the shear layer towards the mean flow. The mechanism of the phenomenon may be interpreted by the impinging jet effect. The spike considered here has a simple conical configuration where the annular jet that expands from the end of the spike has an impinging angle of 30°. Even at present the NPR is slightly below the design point, and the separation bubble is enclosed by the annular main flow producing a higher local pressure up to 1.7 times of the ambient pressure as shown in Fig. 13(e). However, this impinging region will not be present if the spike has been contoured.

At NPR's above 7.75, a new type of shock structure is produced when the internal nozzle starts to operate in a highly under-expanded condition in which an interception shock is generated at the cowl lip while a reflected shock emerges from the spike surface where the internal shock originating from the sharp corner of the throat impinges, as shown in Fig. 12(b). Increasing the NPR to the design point continues to strengthen these shocks. A stronger expansion fan coming from the cowl lip over-expands the flow along the spike surface and leads to the free shock separation formation even in the case of the designed condition shown in Fig. 12(c). For NPR's above 9.87 shown in Fig. 12(d), the pressure on the spike surface is high enough and no flow separation can be captured.

It is important to note here that unlike a contoured spike configuration in a conventional aerospike nozzle, in which the nozzle achieves expansion to the ambient condition by means of a centered expansion fan generated at the cowl lip and terminated at the spike end at the design condition, the expansion fan hits at shorter distances on the spike, and as a consequence, a series of expansion/compression waves continues till the spike end. In the present case, the non-uniform flow exhausting from the internal nozzle

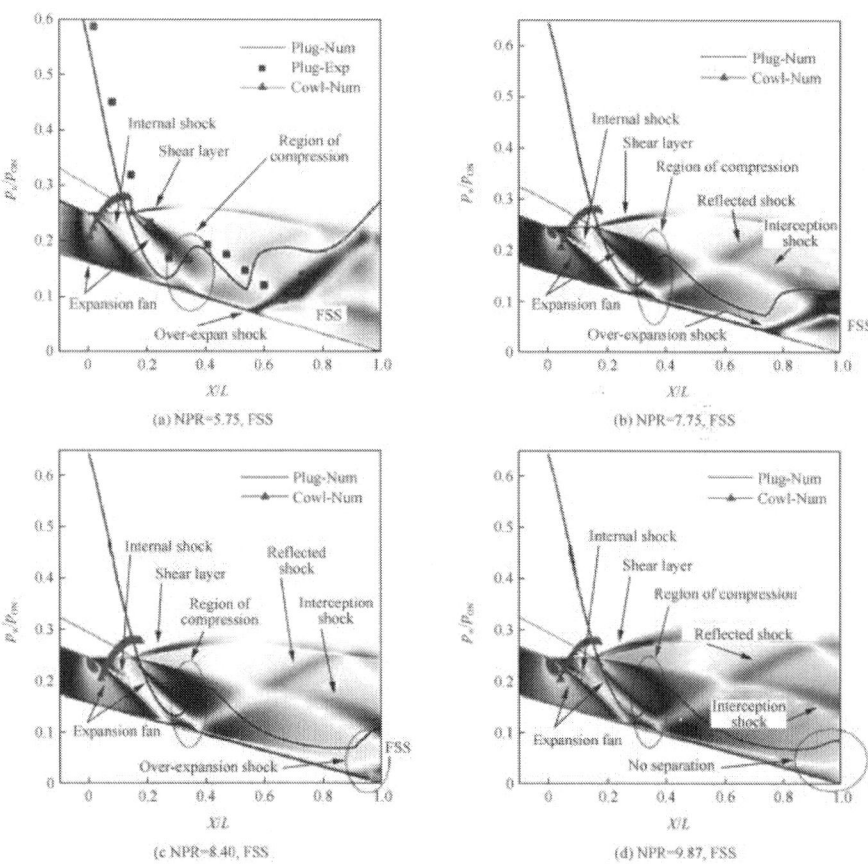

Figure 12. Shock structure motion as the NPR is increased from 5.75 to 9.87 at the sea-level atmospheric conditions, Case 1-1, p_a = 101325 Pa.

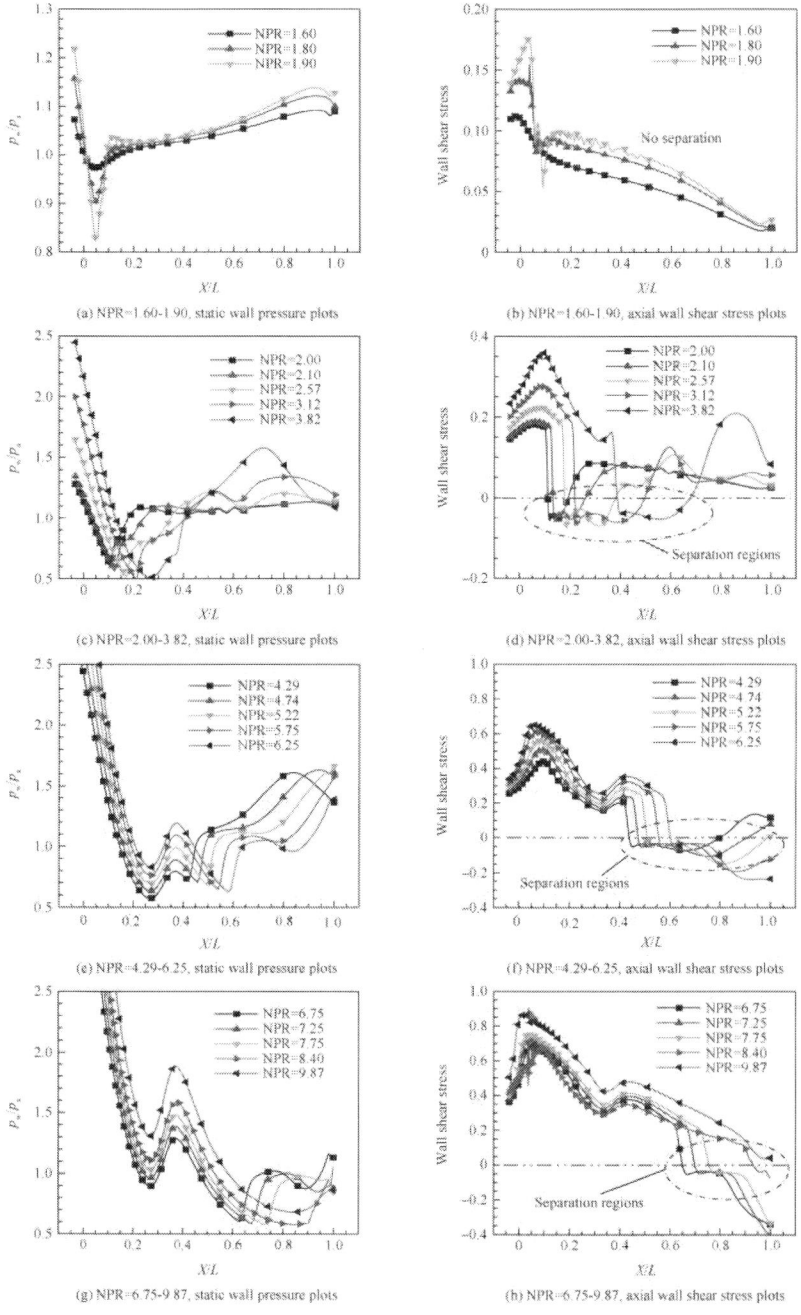

Figure 13. Streamwise distribution of the static wall pressure and the axial wall shear stress along the spike surface for different NPRs in the sea-level atmospheric conditions, Case 1-1, p_a = 101325 Pa.

in addition to the conical configuration of the spike surface makes the spike geometry incapable of canceling out all the impinging waves. Firstly, the flow at the exit of the internal nozzle is not uniform, due to the asymmetric expansion fans coming from the nozzle throat and the following compression shock generated by the sharp change in slope at the throat section. Secondly, when the internal nozzle starts to operate in under-expanded conditions, additional asymmetric expansion fans and interception shock are generated from the cowl lip. Finally, the asymmetric flow passage in addition to that impinging jet effect induces a higher back pressure. The result of these phenomena is the formation of stronger compression shock and sustaining over-expansion of the flow along the spike surface to a shock-induced flow separation even at the design condition.

Static wall pressure and axial wall shear stress profiles

The numerical schlieren images of shock structures constructed from computational results obtain additional shock physics and associated flow separation behavior at various NPR's against the experimental methodology. In particular, the presence of a restricted shock separation regime at low and moderate NPRs while a free shock separation regime at high NPR conditions is recognized. In order to further confirm these flow characteristics, the static wall pressure as well as the axial wall shear stress distribution on the spike is instructive. The plot of the axial wall shear stress on the spike can clearly show the separation as well as the reattachment point induced by the incident and separation shocks by that the value decreasing to below zero indicates that the boundary separates at the present point while increasing to above zero indicates that the boundary reattaches again. Fig. 13 shows the plots of the predicted static wall pressure and the axial wall shear stress distribution on the spike. As the nozzle exhibits a multitude of shock structures and associated flow separation regimes at various NPR's, the figures show a progression of NPRs for the entire range. The throat is located at $X/L = 0$, while the static wall pressure, the axial wall shear stress, and the X co-ordinate have been normalized by the ambient pressure, the maximum value of the axial wall shear stress, and the spike length, respectively.

The plots in Fig. 13(a) and (b) correspond to that of no flow separation regime for NPR's below 1.96, and the flow expands from a subsonic condition at an NPR of 1.60 to a normal shock condition at an NPR of 1.90. Looking at the plot with an NPR of 1.90 for the spike, we notice a bump in the static wall pressure at $X/L = 0.09$, for which as in our earlier discussion, this pressure increase is induced by the normal shock which is not strong enough to induce the flow boundary separation, and it is also evident in the axial wall shear stress distribution that the value stays above and beyond zero along the entire spike as shown in Fig. 13(b). Another feature that can be discerned in the pressure distribution is the pressure adaptation to the ambient condition, in which the spike is enclosed by the annular jet and induces a back pressure higher than that in the ambient condition, and then the static wall pressure increases above and beyond the ambient pressure just aft in the throat region monotonically. This scenario is absent in a conventional bell-type nozzle during transonic conditions.

Fig. 13(c) shows the static wall pressure distribution in the restricted shock separation regime for NPR = 2.00, 2.10, 2.57, 3.12, 3.82. It is evident from the axial wall shear stress plots that the nozzle really exhibits an RSS regime at these NPRs, and in fact the details of the flow pattern cannot be recognized by the experimental methodology.[4] More than five types of shock structures are captured at these NPRs, however, only the moderate NPR = 2.57, 3.12, exhibit a conventional RSS regime while the lower, NPR = 2.00, 2.10, and higher, NPR = 3.82, NPRs are not classical RSS conditions as seen earlier. In case of the conventional RSS regime, the shock exhibits a λ structure and the pressure rises after the first leg of the λ-shock and then forms a plateau which ends when the reflected shock, the second leg of the λ-shock, hits the recirculation bubble. Immediately after this, the pressure increases till the reattachment point and then monotonically decreases as the flow expands downstream from reattachment. However, using the present case of NPR = 3.82 as an example, the expansion fan emerging from the cowl lip over-expands the flow along the spike surface and results in forming of a single oblique over-expansion shock; the pressure rises after the separation shock ($X/L = 0.4$) and continues to increase till the reattachment point ($X/L = 0.66$), and then decreases downstream from reattachment, while no plateau can be discerned along the recirculation bubble.

Fig. 13(e) shows the static wall pressure distribution progression during the flow separation pattern transition regime for NPRs of 4.29–6.25 when the shock structure transforms from the restricted shock separation regime to the free shock separation regime. The transition happens at an NPR of 5.22 as seen earlier, and is also evident in the axial wall shear stress plots shown in Fig. 13(f). In contrast to flow separation transition phenomena in a bell-type nozzle, there is no distinct step change in the evolution of the static wall pressure on the spike. This difference is not entirely clear but is consistent with our computation upon the grid density effect study later. Examination of the static wall pressure plots shows a region of compression (seen as a hump) starting from X/L of 0.26–0.27. The hump grows up with increasing the nozzle pressure ratio while its starting point has slight excursion. In addition, the static wall pressure profile shows similar evolution as the one for a classical RSS condition that the pressure rises after the separation shock and then forms a plateau which ends in a distance, and immediately after this, the pressure increases till p_w/p_a is over 1.5. Moreover, this scenario indicates a different free shock separation behavior from the one in a conventional axisymmetric bell-type nozzle.

In the final figures of the series, Fig. 13(g) and (h), these NPR's shown exhibit that the flow condition on the spike transforms from the free shock separation at NPRs of 1.60–8.40 to full expansion at an NPR of 9.87 although the static wall pressure at the end of the spike surface is still below that in the ambient condition. The plots of the axial wall shear stress show clearly the progression of the separation point. The region of compression on the spike now shows a distinct static wall pressure plot hump. By using both the static wall pressure non-dimensionalized with respect to the ambient condition and the axial wall shear stress profile, the fact that the expansion fan coming from the cowl lip over-expands the flow along the spike surface and leads to the free shock separation formation even in the case of the designed condition can be clearly recognized.

Prediction of the flow separation behavior

As flow separation may lead to performance losses and undesired high nozzle structural loads,[47, 48 and 49] an accurate separation criterion is crucial. A series of separation criterion has been developed with increasing knowledge and availability of experimental data. However, most of the historical data are predominantly for conical or bell-type nozzles with an axisymmetric configuration, and thereby applicability on the present asymmetric nozzle needs to be validated. Fig. 14 plots the separation data (p_{sep}/p_a or p_{sep}/p_{ON}) as a function of the corresponding wall separation Mach number, Ma_{sep}, or the NPR, on the spike surface as well as the cowl wall, where the wall separation Mach number is based on the isentropic ratio of p_{ON}/p_{sep}. Three separation criteria are plotted for comparison: the well-known Schmucker criterion [50] (1), the separation criterion (2) for turbulent nozzle flows suggested by Stark et al. [51 and 52], and the separation criterion (3) suggested by Ge et al. [53] recently based on flow separation data in asymmetric ramp nozzles.

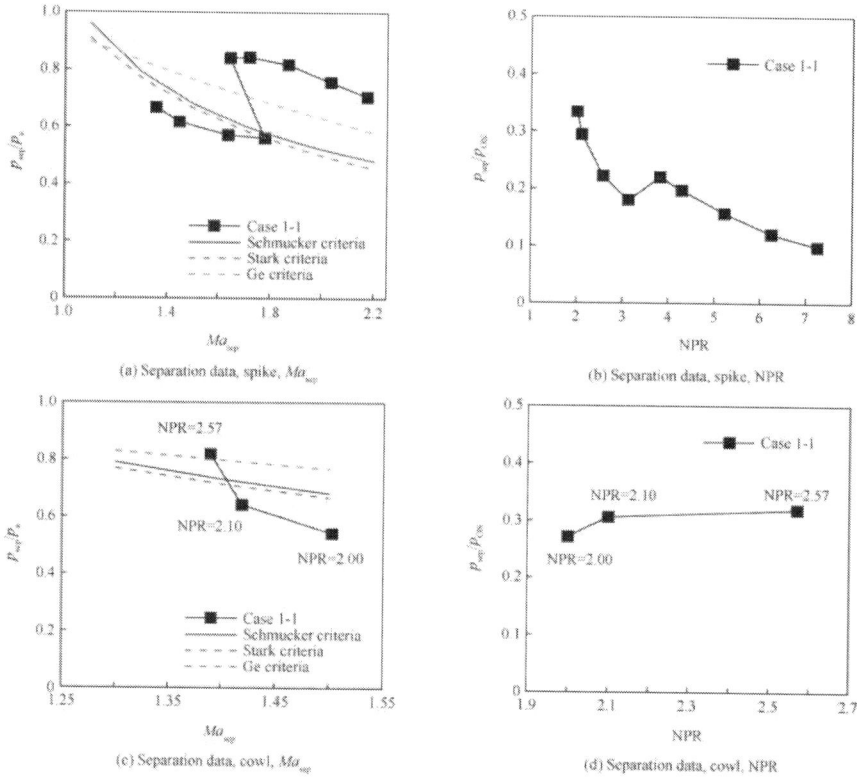

Figure 14. Comparison of separation data plots on the spike and cowl surfaces at the sea-level atmospheric conditions, Case 1-1, $p_a = 101325$ Pa as a function of Ma_{sep} and NPR, respectively.

$$\frac{p_{sep}}{p_a} = \left(1.88 Ma_{sep} - 1\right)^{-0.64} \tag{1}$$

$$\frac{p_{sep}}{p_a} = \frac{1}{Ma_{sep}} \tag{2}$$

$$\frac{p_{sep}}{p_a} = -1.76\left(0.47 Ma_{sep}^{0.45} - 1\right) \tag{3}$$

Looking at the plots of separation data on the spike wall shown in Fig. 14(a) and (b), the data affected by the region of compression, the over-expansion by the expansion fan generated at the cowl lip, and the enclosed back pressure environment by the annular jet for NPR $\geqslant 3.82$ when the internal nozzle starts to operate in an under-expanded condition are clearly pointed out in the plots. A region covers nozzle pressure ratios of 3.12–3.82, corresponding to the wall separation Mach number of 1.65–1.80, and divides the data-set into two regions. The data show a strong variation and are bounded below by all the three separation criteria for NPR $\leqslant$ 3.12 and above by the criteria for NPR $\geqslant$ 3.82. At first sight, an important separation behavior transformation seems to happen in between as discussed earlier. The data of Schmucker and Stark criteria show below NPR $\leqslant$ 3.12 a trend which seems to be a sight parallel shift and reproduces the separation pressure with a reasonable accuracy for this NPR regime, while failing at the prediction for the NPR $\geqslant$ 3.82. The separation criterion (3) suggested by Ge et al. fails to reproduce the separation pressure data for both of the two regions. This is a significant hint that the asymmetric nozzle configuration may affect the flow separation behavior in the present aerospike nozzle and induce the miss-prediction by the separation criteria developed from separation data in conical or bell-type rocket nozzles.

Some interesting effects are also pointed out in the plots of separation data on the cowl wall shown in Fig. 14(c) and (d). For NPR's between 2.00 and 2.57, the flow separation on the cowl wall is always in the FSS regime as seen earlier. However, with the interception effect of the oblique compression shock induced by the separated shear layer at the sharp corner, the flow separation on the cowl surface exhibits abnormal free shock separation behavior. The separation pressure, p_{sep}, show an opposite trend as a function of NPR that increasing the NPR strengthens the upstream oblique compression shock, leading to a higher local static wall pressure and total pressure loss, and then to a lower wall separation Mach number, Ma_{sep}. Again, all the three separation criteria fail to reproduce the separation data on the cowl wall.

Gas density effect

The above discussion gives a detailed view of the shock physics and flow separation behavior as the NPR is increased from low to high values for the sea-level condition. Now, we compare the gas density effect on the above-mentioned flow characteristics at specific NPRs for the three ambient pressure conditions. Recent studies[54 and 55] on a thrust-optimized parabolic nozzle have reported significant variations in results when comparing data from tests conducted in sea-level atmosphere with those inside a high-altitude chamber. The series of pictures illustrated in Fig. 15 shows the predicted shock patterns at the three different ambient pressure conditions for NPR = 2.00, 2.10, and 2.57, when more complex shock/boundary layer interaction is expected for these highly over-expanded conditions. It is noteworthy here that the results of shock-induced flow separation structures and location as well as the static wall pressure profiles variations with the three different ambient pressure conditions show a similar tendency as those for NPR = 2.57. As a consequence, the results for higher NPRs are not included in the paper. Pictures are also contoured with the same levels by the data constructed from the computational results with the absolute values of the first gradients of density at the grid nodes. At first sight, the shock structure is somewhat less distinct in the high-altitude simulations, indicating a weaker shock system for lower density flows.

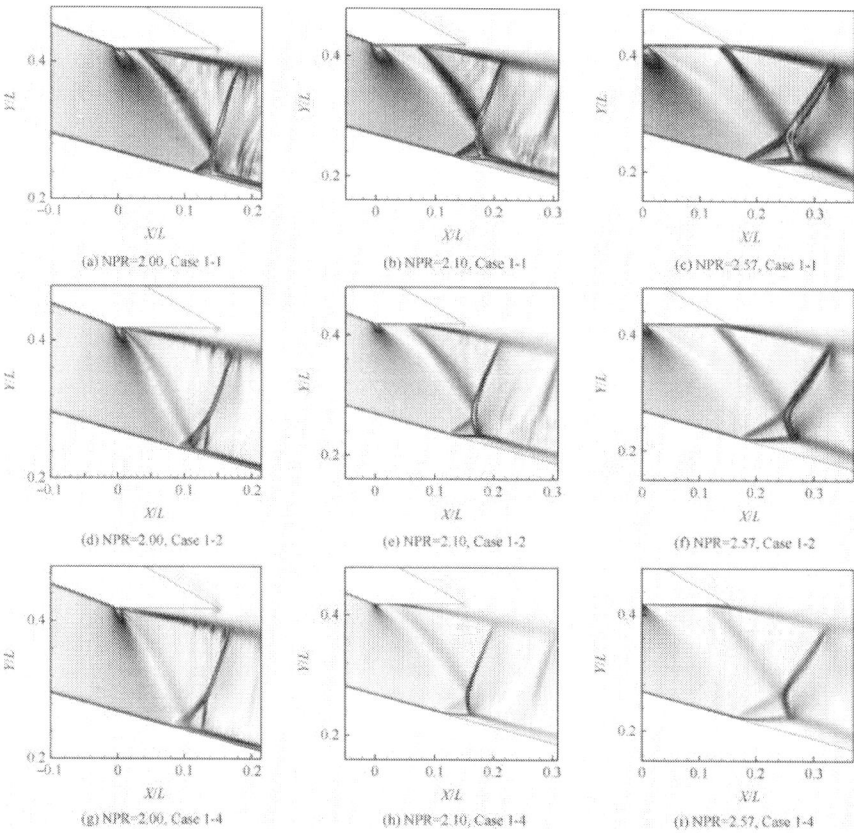

Figure 15. Comparison of the predicted shock structures at different ambient pressure conditions for NPR = 2.00, 2.10, and 2.57. Pictures are contoured at the same levels by the data constructed from the computational results with the absolute values of the first gradients of density at the grid nodes.

Looking at the pictures on the left of Fig. 15(a), (d), and (g) from top to bottom for an NPR of 2.00, we note that the predicted shock patterns and associated flow separation on both the spike and cowl surfaces in the two high-altitude simulations exhibit distinct differences in contrast to the sea-level condition, suggesting a strong gas density effect at this NPR condition. As discussed earlier, this particular NPR of 2.00 lies in the shock transition from a normal shock to Mach reflection on the spike surface while the separation point jumps from the throat's sharp corner to the downstream cowl extension. For flows in the high-altitude simulations, the shock structure generates a λ pattern enclosing a smaller recirculation zone at the spike. The oblique compression shock starting from the throat's sharp corner is weak now and hits on the first foot of the lambda shock; the separation shock is modified in the vicinity of the wall by the intersection between these shocks.

In terms of cowl side, decreasing the ambient pressure pushes the separation point on the cowl surface back to the throat's sharp corner, and it freezes exactly when the ambient pressure increases from 25% to 50% at high-altitude simulation conditions, suggesting that a higher inlet total pressure is needed to push the separation point to jump downstream into the cowl extension for lower density flow. This phenomenon is similar to the gas density effect on the sneak transition process in dual-bell nozzles, and recent studies[54] on a subscale dual-bell nozzle have reported that the separation point for tests inside a high-altitude simulation chamber moves into the region of wall inflection much earlier and stays there for a much longer time. This delays the process of transition and hence increases the NPR of dual-bell transition as p_{ON} is decreased for tests inside the high-altitude simulation chamber. For results at higher nozzle pressure ratios of 2.10 and 2.57 shown in the middle and right list of Fig. 15, the comparison between the predicted shock patterns at the three ambient conditions shows a close one at these higher NPRs. One difference between the three results is that the separation point shows a distinct excursion upstream on both the spike and cowl surfaces as the ambient pressure decreases. The other feature that can be discerned is a smaller excursion of the separation point at a higher NPR of 2.57 than that observed for an NPR of 2.10. This suggests that the gas density effect may be weakening at a higher Mach number flow regime, when the flow Reynolds number increases with the increase of nozzle pressure ratio as well as flow Mach number, and the Reynolds number has been reported to have a significant influence on the shock/boundary layer interaction. [54]

Fig. 16(a)–(c) plot the streamwise distribution of the non-dimensional mean static wall pressure for the three NPRs at different ambient conditions. Again, the plot of the axial wall shear stress along the spike is used here to interpret the excursion behavior of the separation shock as shown in Fig. 16(d)–(f). It may be noted that approximately similar values of the static wall pressure are being experienced irrespective of the simulated ambient pressure conditions except in two regions. One difference is that the lowest pressure value just before the separation shows an increase with a decrease in the ambient pressure. Therefore, a lower pressure rise is needed for the static wall pressure adaptation to the ambient condition, resulting in a weaker separation shock for the lower-density flow, as shown in the predicted shock patterns. The second difference is the absent wavy plots of the static wall pressure distribution for the two high-altitude simulation cases. This suggests that, for lower gas density, the weak expansion and compression waves in the aft region of the main normal shock are also weakened by the separation shock and even disappear for extremely low ambient pressure conditions. A distinct decrease of the axial wall shear stress with a decrease in the ambient pressure is shown in the axial wall shear stress plots, Fig. 16(d)–(f). The excursion upstream from the separation point location, the reduction in the lateral extent of the separation zone, and the weaker influence on the above two features with a decrease in the ambient pressure are shown clearly in the pictures.

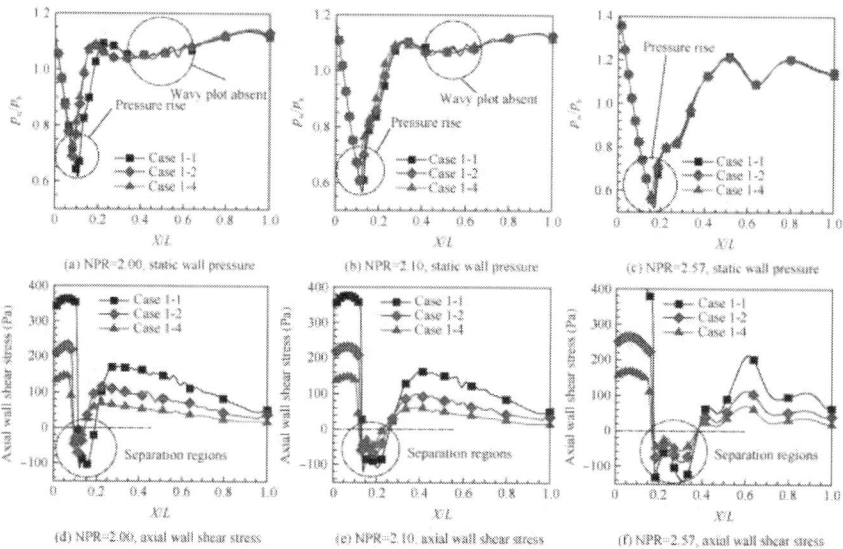

Figure 16. Comparison of the static wall pressure and the axial wall shear stress distribution along the spike wall at different ambient pressure conditions for NPR = 2.00, 2.10, and 2.57.

Fig. 17 shows the variation of the streamwise distribution of the non-dimensional mean static wall pressure and the axial wall shear stress along the cowl surface. Because the length of the cowl, l, is smaller compared with that of the spike, the gas density effect is more distinct for the separation behavior on the cowl surface. It can be noted that the static wall pressure distribution shows a larger discrepancy in the separation region, while with a decrease in the ambient pressure, the rate of pressure rise across the separation shock decreases gradually. This feature is less distinct with the increase of the NPR shown in Fig. 17(c). The phenomenon that the location of the incipient separation point excurses back to the throat's sharp corner and freezes exactly even with large changes in the ambient pressure (transition from the 25% to the 50% ambient pressure case) is shown clearly in the axial wall shear stress plots, Fig. 17(d)–(f).

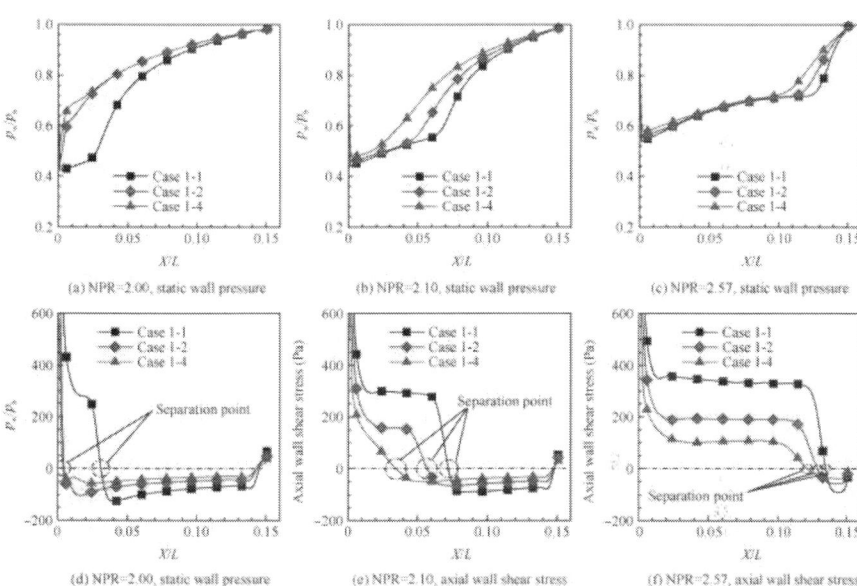

Figure 17. Comparison of the static wall pressure and the axial wall shear stress distribution along the cowl wall at different ambient pressure conditions for NPR = 2.00, 2.10, and 2.57.

To investigate the reason behind the observed discrepancies, Fig. 18 shows a comparison of the static wall pressure gradient near the separation shock region for the three ambient conditions at NPR = 2.10, 2.57. Here the static wall pressure gradient, dp/dx, is non-dimensional (the static wall pressure, p_w, is non-dimensionalized with respect to the ambient pressure, p_b) and the X co-ordinate is normalized by the annular gap at the throat section h_t. The separation point locates between X/h_t of 0.5–1.0 and 1.0–1.5 for NPR = 2.10, 2.57, respectively. It can be seen that approximately similar values of dp/dx are being experienced outside the separation region irrespective of the simulated gas density. However, as the ambient pressure increases, the value of dp/dx shows a decrease with a decrease in gas density across the separation shock. Studies on bell-type nozzles have also indicated that a distinct excursion of the incipient separation point occurs when the adverse pressure gradient across the separation shock is smaller. [56] This suggests that dp/dx is larger for higher gas density and vice versa, indicating a stronger shock system for higher density flow.

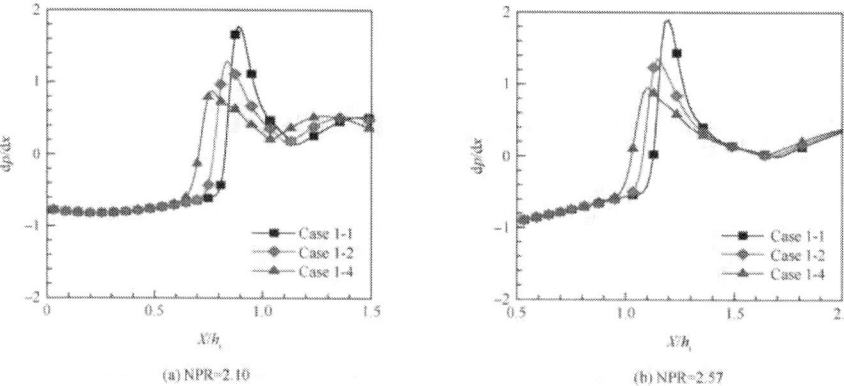

(a) NPR=2.10   (b) NPR=2.57

Figure 18. Comparison of the static wall pressure gradient across separation at different ambient pressure conditions for NPR = 2.10, 2.57.

During a low-altitude mode with NPR = 2.00, the shock patterns and associated flow separation behavior strongly signify the compression shock and the separated shear layer starting from the throat's sharp corner. As discussed earlier, once the throat region has reached the supersonic regime, the wall inflection at the sharp corner that controls separation occurs over a wide range of NPRs because a large pressure gradient prevalent at the corner allows only a small movement of the separation point with large changes in the NPR. In particular, the gas density effect delays this process because of lower dp/dx. Additional calculations and experimental tests on high-altitude simulation are needed to get deeper insight into the aerodynamic mechanism behind the observed gas density effect.

SUMMARY AND CONCLUSIONS

(1) The computational results agree well with the experimental data in both shock physics and static wall pressure distribution, indicating that

the axisymmetric computational methodology here is advisable to accurately predict the flow physics. The progressively increased excursion in the plots of static wall pressure distribution with an increase in the NPR may contribute to slightly miss-mimicking the experimental condition because of the jet entrainment effect.

(2) The annular conical aerospike nozzle is observed to be dominated by shock/shock and shock/boundary layer interactions at all calculated NPRs, and the shock physics and associated flow separation behavior are quite complex. Increasing the NPR changes the operating condition of the internal nozzle as well as the basic flow physics such as the expansion fans, the separated shear layer, and the over-expansion shock on both the spike and cowl surfaces. As the internal nozzle operates from the highly over-expanded to the over-expanded condition and then to the under-expanded condition, the flow condition on the spike exhibits a normal shock structure with no flow separation for NPR < 2.0, a multitude of shock structure transitions with the restricted shock separation for $2.0 \leqslant$ NPR $\leqslant 5.22$, and an oblique over-expansion shock with the free shock separation for $5.22 <$ NPR $\leqslant 8.40$ regimes. This shock and separation transitions are absent in optimized bell-type nozzles nor the reported shrouded plug nozzle, and the identification of these three regimes helps in further investigation on the unsteady fluid dynamics and related side loads generation. The computational results provide additional flow characteristics against the experimental data; in particular, the shock/shock and shock/boundary layer interactions with the restricted shock separation at highly over-expanded conditions and the free shock separation behavior at higher NPRs than those at design conditions are excluded in the experimental study.

(3) The separation data show that a strong variation for the spike, a region that covers NPR of 3.12–3.82, corresponding to the wall separation Mach numbers of 1.65–1.80, divides the data-set into two regions. All the three separation criterias fail to reproduce the separation data of the spike for the NPR $\geqslant 3.82$, when the internal nozzle starts to operate in under-expanded conditions. On the other hand, the flow separation structure on the cowl surface is always in the FSS regime. However, with the interception effect of the oblique compression shock induced by the separated shear layer at the sharp corner, the separation pressure, p_{sep}, shows an opposite trend as a function of the NPR in contrast to that in over-expanded conventional bell-type rocket nozzles. We conclude that the separation criteria developed from separation data in conical or bell-type rocket nozzles may be inapplicable for the prediction of flow separation behavior in the present asymmetric supersonic nozzle.

(4) A strong gas density effect has been found at the highly over-expanded condition with NPR = 2.00, when the wall inflection at the sharp corner that controls separation occurs over a wide range of NPRs during this flow regime. However, for results at higher NPRs, the comparison between the predicted shock patterns at the three ambient conditions shows a close one for these higher NPRs regime. This suggests that the gas density effect is going to weaken at a higher Mach number flow regime when the flow Reynolds number increases with an increase of the NPR as well as the flow Mach number. The adverse pressure gradient across the separation shock, dp/dx, is larger for higher gas density and vice versa, indicating a stronger shock system for higher density flow, which may contribute to the observed gas density effect. These results emphasize that the flow separation behavior tests in such an aerospike nozzle inside a high-altitude test facility should be carefully interpreted; in particular, Reynolds number effects are needed to be concerned when comparing results from the sea-level conditions.

REFERENCES

1. Fick M, Schmucker RH. Performance aspects of plug cluster nozzles. J Spacecraft Rockets 1996;33(4):507–12.
2. Rommel T, Hagemann G, Schley CA, Krulle G, Manski D. Plug nozzle flowfield analysis. J Propul Power 1997;13(5):629–34.
3. Hagemann G, Immich H, Terhadt M. Flow phenomenon in advanced rocket nozzles-the plug nozzle. Reston: AIAA; 1998. Report No.: AIAA-1998-3522.
4. Verma SB. Performance characteristics of an annular conical aerospike nozzle with freestream effect. Reston: AIAA; 2008. Report No.: AIAA-2008-5290.
5. Schwane R, Hagemann G, Reijasse P. Plug nozzles-assessment of prediction methods for flow features and engine performance. Reston: AIAA; 2002. Report No.: AIAA-2002-0585.
6. Onofri M. Pug nozzles: summary of flow features and engine performance. Reston: AIAA; 2002. Report No.: AIAA-2002-0584.
7. Kapilavai DSK, Tapee J, Sullivan J, Merkle CL, Wayman TR, Conners TR. Experimental testing and numerical simulations of shrouded plug-nozzle flowfields. J Propul Power 2012;28(3): 530–44.
8. Arens M, Spiegler E. Shock-induced boundary layer separation in overexpanded conical exhaust nozzles. AIAA J 1963;1(3):578–81.

9. Schmucker RH. Flow processes in overexpanded chemical rocket nozzles. Part 2: side loads due to asymmetric separation. Washington, D.C.: NASA; 1984. Report No.: NASA-TM-77395.

10. Nave LH, Coffey GA. Sea level side loads in high-area-ratio rocket engines. Reston: AIAA; 1973. Report No.: AIAA-1973- 1284.

11. Nguyen AT, Deniau H, Girard S, de Roquefort TA. Unsteadiness of flow separation and end-effects regime in a thrust optimized contour rocket nozzle. Flow Turbul Combust 2003;71(1–4):161–81.

12. Hagemann G, Frey M, Koschel W. Appearance of restricted shock separation in rocket nozzles. J Propul Power 2002;18(3): 577–84.

13. Ostlund J. Flow processes in rocket engine nozzles with focus on flow-separation and side-loads. Mekanik (Stockholm): Royal Institute of Technology; 2002.

14. Chen CL, Chakravarthy SR, Hung CM. Numerical investigation of separated nozzle flows. AIAA J 1994;32(9):1836–43.

15. Gross A, Weiland C. Numerical simulation of separated cold gas nozzle flows. J Propul Power 2004;20(3):509–19.

16. Deck S, Nguyen AT. Unsteady side loads in a thrust-optimized contour nozzle at hysteresis regime. AIAA J 2004;42(9):1878–88.

17. Nasuti F, Onofri M. Viscous and inviscid vortex generation during start-up of rocket nozzles. AIAA J 1998;36(5):809–15.

18. Morı´nigo JA, Salva´ JJ. Three-dimensional simulation of the selfoscillating flow and side-loads in an over-expanded subscale rocket nozzle. Proc IMechE Part G J Aerosp Eng 2006;220(5):507–23.

19. Baars WJ, Tinney CE. Transient wall pressures in an overexpanded and large area ratio nozzle. Exp Fluids 2013;54(2):1–17.

20. Baars WJ, Tinney CE, Ruf JH, Brown AM, McDaniels DM. Wall pressure unsteadiness and side loads in overexpanded rocket nozzles. AIAA J 2012;50(1):61–73.

21. Ostlund J, Damgaard T, Frey M. Side-loads phenomena in highly over-expanded rocket nozzles. Reston: AIAA; 2001. Report No.: AIAA-2001-3684.

22. Stark R, Kwan W, Quessard F, Hagemann G, Terhardt M. Rocket nozzle cold gas test campaigns for plume investigations. Proceeding of the 4th European symposium on aerothermodynamics for space vehicles; 2001 Oct 15-18; Capua, Italy. ESA SP-487, 2002

23. Tomita T, Sakamoto H, Onodera T, Sasaki M, Takahashi M, Watanabe Y. Experimental evaluation of side-loads characteristics on TP, CTP and TO nozzles. Reston: AIAA; 2004. Report No.:AIAA-2004-3678.

24. Mouronval AS, Hadjadj A. Numerical study of the starting process in a supersonic nozzle. J Propul Power 2005;21(2):374–8.

25. Roquefort de TA. Unsteadiness and side-loads in over-expanded nozzles. Proceeding of the 4th European symposium on aerothermodynamics for space vehicles; 2001 Oct 15–18; Capua, Italy. ESA SP-487, 2002.

26. Ostlund J, Muhammad-Klingmann B. Supersonic flow separation with application to rocket engine nozzles. Appl Mech Rev 2005;58(3):143.

27. Page RH, Meyer AP. Hydraulic analog investigation of a plug nozzle. ARS J 1961;31(3):447–8.

28. Angelino G. Approximate method for plug nozzle design. AIAA J 1964;2(10):1834–5.

29. Giel Jr TV, Mueller TJ. Mach disk in truncated plug nozzle flow. AIAA J 1976;13(4):203–7.

30. Berman K, Crimp Jr FW. Performance of plug-type rocket exhaust nozzles. ARS J 1961;31(3):18–23.

31. Rao GVR. Recent developments in rocket nozzle configurations. ARS J 1961;31(3):1488–94.

32. Verma SB, Viji M. Freestream effects on base pressure development of an annular plug nozzle. Shock Waves 2011;21(2):163–71.

33. Verma SB, Viji M. Linear-plug flowfield and base pressure development in freestream flow. J Propul Power 2011;27(6): 1247–58.

34. Ruf JH, McConnaughey PK. The plume physics behind aerospike nozzle altitude compensation and slipstream effect. Reston: AIAA; 1997. Report No.:AIAA-1997-3217.

35. Menter FR, Kuntz M, Langtry R. Ten years of industrial experience with the SST turbulence model. In: Harjalic K, Nagano Y, Tummers M, editors. Turbunlence, heat, and mass transfer 4. New York: Begell House, Inc.; 2003. p. 625–32.

36. Hunter CA. Experimental, theoretical and computational investigation of separated nozzle flows. Reston: AIAA; 1998. Report No.:AIAA-1998-3107.

37. Xiao Q, Tsai HM, Papamoschou D. Numerical investigation of supersonic nozzle flow separation. AIAA J 2007;45(3):532–41.

38. Xiao Q, Tsai HM, Liu F. Computation of turbulent separated nozzle flow by a lag model. J Propul Power 2005;21(2):368–71

39. Xiao Q, Tsai HM, Papamoschou D, Johnson A. Experimental and numerical study of jet mixing from a shock containing nozzle. J Propul Power 2009;25(3):688–96.

40. Carlson JR. A nozzle internal performance prediction method. Hampton (Virginia): Langley Research Center; 1992. Report No.: NASA-TM-3221.

41. Wang TS. Transient three-dimensional startup side load analysis of a regeneratively cooled nozzle. Reston: AIAA; 2008. Report No.:AIAA-2008-4300.

42. Sajben M, Kroutil JC. Effect of initial boundary layer thickness on transonic diffuser flow. AIAA J 1983;19(11):1386–93.

43. Henne P. The case for small supersonic civil aircraft. Reston: AIAA; 2003. Report No.:AIAA-2003-2555.

44. Chpoun A, Passerel D, Li H, et al. Reconsideration of oblique shock wave reflections in steady flows. Part 1. Experimental investigation. J Fluid Mech 1995;301:19–35.

45. Coanda H. Device for deflecting a stream of elastic fluid projected into an elastic fluid. United States patent US 2052869. 1936 Sep 1.

46. Kumada M, Mabuchi I, Oyakawa K. Studies on heat transfer to turbulent jets with adjacent boundaries. Bull Jpn Soc Mech Eng 1972;15(88):1246–56.

47. Ostlund J, Damgaard T, Frey M. Side-load phenomena in highly overexpanded rocket nozzles. J Propul Power 2004;20(4): 695–704.

48. Verma SB, Stark R, Haidn O. Relation between shock unsteadiness and the origin of side-loads in a thrust optimized parabolic rocket nozzle. Aerosp Sci Technol 2006;10(6):474–83.

49. Verma SB, Oskar H. Study on restricted shock separation phenomena in rocket nozzles. Reston: AIAA; 2006. Report No.:AIAA-2006-1431.

50. Lawrence RA. Symmetrical and unsymmetrical flow separation in supersonic nozzles dissertation. Dallas: Southern Methodist University; 1967.

51. Stark R. Beitrag zum Versta"ndnis der Stro"mungsablo"sung in Raketendu"sen dissertation. Aachen: RWTH Aachen University; 2010.

52. Stark R, Wagner B. Experimental study of boundary layer separation in truncated ideal contour nozzles. Shock Waves 2009;19(3):185–91.

53. Ge JH, Xu JL, Wang MT, Mo JW. Prediction of flow separation in asymmetric ramp nozzle. Acta Aeronautica et Astronaut Sin 2012;33(8):1394–9 Chinese.

54. Verma SB, Stark R, Haidn O. Reynolds number influence on dualbell transition phenomena. J Propul Power 2013;29(3):602–9.

55. Verma SB, Haidn O. Cold gas testing of thrust-optimized parabolic nozzle in a high-altitude test facility. J Propul Power 2011;27(6):1238–46.

56. Frey M, Hagemann G. Critical assessment of dual-bell nozzles. J Propul Power 1999;15(1):137–43

CITATION

Miaosheng He, Lizi Qin, Yu Liu, Numerical investigation of flow separation behavior in an over-expanded annular conical aerospike nozzle, Chinese Journal of Aeronautics, Volume 28, Issue 4, August 2015, Pages 983-1002, ISSN 1000-9361, http://dx.doi.org/10.1016/j.cja.2015.06.016.

CHAPTER 4

Computational Aerodynamic Analysis on Perimeter Reinforced (Pr)-Compliant Wing

N.I. Ismail[1], A.H. Zulkifli[1], M.Z. Abdullah[2], M. Hisyam Basri[1], Norazharuddin Shah Abdullah[3]

[1] Faculty of Mechanical Engineering, Universiti Teknologi MARA, 40450 Shah Alam, Selangor, Malaysia

[2] School of Mechanical Engineering, Universiti Sains Malaysia, Engineering Campus, 14300 Nibong Tebal, Penang, Malaysia

[3] School of Materials and Mineral Resources Engineering, Universiti Sains Malaysia, Engineering Campus, 14300 Nibong Tebal, Penang, Malaysia

ABSTRACT

Implementing the morphing technique on a micro air vehicle (MAV) wing is a very challenging task, due to the MAV's wing size limitation and the complex morphing mechanism. As a result, understanding aerodynamic characteristics and flow configurations, subject to wing structure deformation of a morphing wing MAV has remained obstructed. Thus, this paper presents the investigation of structural deformation, aerodynamics performance and flow formation on a proposed twist morphing MAV wing design named perimeter reinforced (PR)-compliant wing. The numerical simulation of two-way fluid structure interaction (FSI) investigation consist of a quasi-static aeroelastic structural analysis coupled with 3D incompressible Reynolds-averaged Navier–Stokes and shear-stress-transport (RANS–SST) solver utilized throughout this study. Verification of numerical method on a rigid rectangular wing achieves a good correlation with available experimental results. A comparative aeroelastic study between PR-compliant to PR and rigid wing performance is organized to elucidate the morphing wing performances. Structural deformation results show that PR-compliant wing is able to alter the wing's geometric twist characteristic, which has directly influenced both the overall aerodynamic performance and flow structure behavior. Despite the superior lift performance result, PR-compliant wing also suffers from massive drag penalty, which has consequently affected the wing efficiency in general. Based on vortices investigation, the results reveal the connection between these aerodynamic performances with vortices formation on PR-compliant wing.

INTRODUCTION

Micro air vehicle (MAV) is a small aircraft with a wingspan of less than 15 cm, flying at low Reynolds number regime (10^4–10^5). It is an alternative tool to replace unmanned aerial vehicles (UAVs) for intelligence and surveillance in confined space areas. Various missions, including in the battlefield have presented a huge potential for MAV use. The rigid wing MAV type is a popular choice among researchers since it offers better payload and endurance capability.[1] However, the rigid wing MAV also suffers a few drawbacks such as flow separation bubbles, large wing-tip vortex swirling,[2] difficult flight controllability[3] and small center of gravity (CG) range.[4] In this case, the biological inspirations design have suggested the solution idea (to the drawbacks) by introducing passive (known as membrane wing design)[5] and [6] and active shape adaptation (known as morphing wing design).[7] Most of these biological designs are inspired from flying characteristics of airborne mammals,[8] birds[9] and insects.[10]

Technically, morphing wing is a method where the wing changes its shape during flight in order to optimize the aerodynamic performance.[11] Planform deviation through wingspan alteration, chord length change and swept angle variation can be defined as morphing. Morphing can also be produced through chordwise or spanwise wing bending.[12] Twist morphing method has been used as a practical control technique in flight dynamics[7] and can improve aerodynamic properties for bigger aircraft.[13] However, morphing wing design is also well associated with complicated mechanism. Mujahid[14] had to use a space consuming combination of servos, torque tubes and linkages to perform the morphing wing technique in his MAV design. This might be the reason why his MAV wingspan had to be extended to more than 10 inches, which obviously exceeds the MAV definition of less than 15 cm (6 inches) wingspan size. Therefore, a proper design of MAV wing with compliant mechanism structure is much desired, since it can significantly reduce weight, offers more simplicity and potentially better aircraft performance.[15]

Performing a morphing technique on a MAV-sized wing is a very challenging design task, due to the MAV's wing size limitation and a possibly complex morphing mechanism.[16] As a result, understanding the aerodynamic characteristics and flow structures subject to wing structure deformation of a morphing MAV wing has remained obstructed. Thus, this present work is carried out to investigate the aerodynamics performance and flow structures behavior over a proposed morphing wing design (PR-compliant wing) subject to its structural deformation. The PR-compliant wing is designed based on standard PR wing combined with a compliant mechanism component. A quasi-static aeroelastic study by using the fluid structure interaction (FSI) method is utilized to elucidate the performance enhancement of this proposed morphing wing design.

FSI COMPUTATION METHOD

Governing equation

Fluid solver
Steady, incompressible, turbulent flow boundary conditions are utilized in this present study, in which, the airflow field is solved based on RANS equations coupled with SST $k-\omega$ turbulent model. In addition, the SST model is further refined by employing a blending of the turbulent eddy viscosity formulation. [17]

Structural solver
The governing equation for a continuum undergoing motion of steady deformation in this case is solved based on the Cauchy's equation:

$$\nabla \sigma ij + f = 0 \qquad (1)$$

where σ_{ij} is the stress tensor and f the external force component ($i, j = 1, 2, 3$ for 3D structures). [17] The force components that are applied in this present case, involve the external force, the fluid pressure and shear at the fluid–solid boundary.

ANSYS FSI computational framework
Arbitrary Lagrangian–Eulerian (ALE) description is employed in CFX FSI computation. The FSI is achieved by satisfying either a velocity or displacement continuity, respectively, at fluid and solid boundaries, Γ. In this FSI computation, the force equilibrium is achieved at the boundary interface between both fluid and solid domain. The details of this FSI theory can be found in Ansys literature.[17] The strongly coupled FSI simulation process is summarized in Fig. 1.

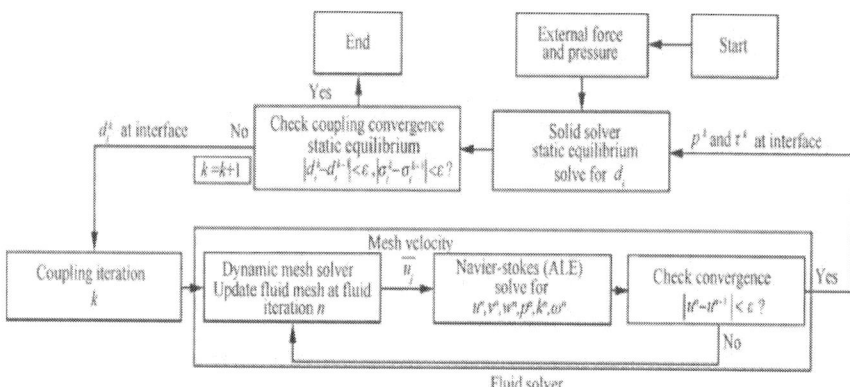

Figure 1. FSI simulation process.

MAV wing model and boundary conditions

The MAV wing model

The MAV wing models that used in the present research are rectangular, PR-compliant, PR and rigid wings. The rigid rectangular wing model is used for numerical verification studies, to justify and ascertain the method's validity in terms of pre-processing setup and boundary condition selections. The verification is accomplished by comparing the numerical results of rectangular wing configuration with available experimental data from Mueller and Torres[18]. PR and rigid wing baseline design are purposely included in the comparative study in order to elucidate the enhancement of PR-compliant wing performance.

The PR-compliant wing (as depicted in Fig. 2), meanwhile, is a proposed morphing wing model, that is almost identical to the membrane wing for MAVs, developed by researchers at the University of Florida (UF),[5] and [6] with the fuselage, propeller, and stabilizers removed. It is conceded that propellers may play a part in MAV aerodynamic performance, but the mechanics and dynamics in this case (no propeller with membrane wing) are still not fully understood or documented in MAV studies.[19] Furthermore, the wing shape and dimensions used in this study are nearly the same as reported by Refs.[5] and [6] that removed propellers from their respective studies for simplification purposes.

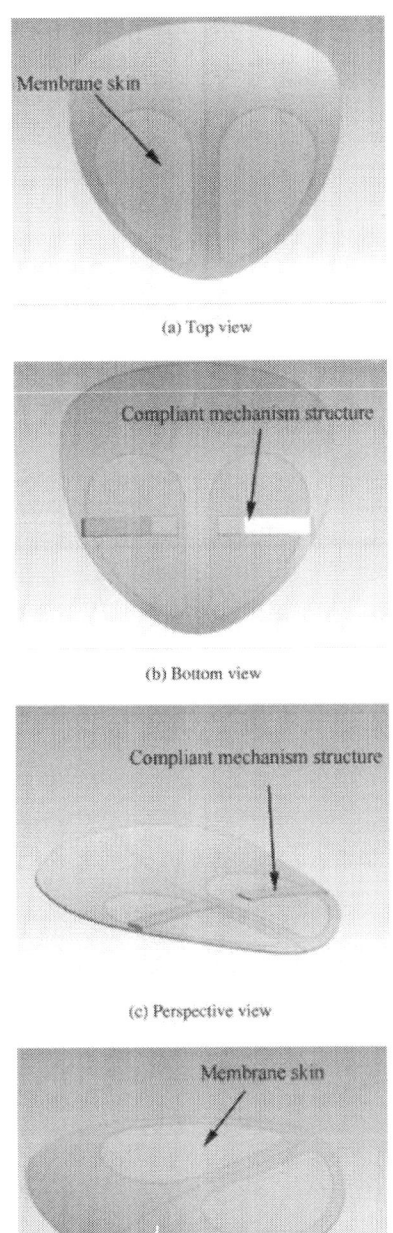

(a) Top view

(b) Bottom view

(c) Perspective view

(d) Isometric view

Figure 2. PR-compliant wing.

The PR-compliant wing structure has an integrated compliant mechanism (which was adopted from previous works of Palmisano et al.[20]), attached at the lower wing surface in order to activate the twist morphing actuation. The physical structure and basic kinematic principle of this compliant mechanism are shown in Fig. 3. The compliant mechanism is needed simply to produce a y-direction displacement at the wingtip. The wingtip displacement magnitude is considered an output, with the force actuation level being an input. Basically, the objective function of this compliant mechanism is to maximize the wingtip y-direction displacement (downward deformation/deflection) under a specified input force. Based on structural optimization study, the compliant mechanism structure is located at a position 90 mm from the leading edge and perpendicular to the wing chordwise axis. The gap between the compliant mechanism components is intentionally permitted for future works on force generator devices. The main characteristics of PR wing such as shape planform, camber location, wing twist and dihedral angle have been altered from original UF wing due to finite element modeling simplification and additional works.

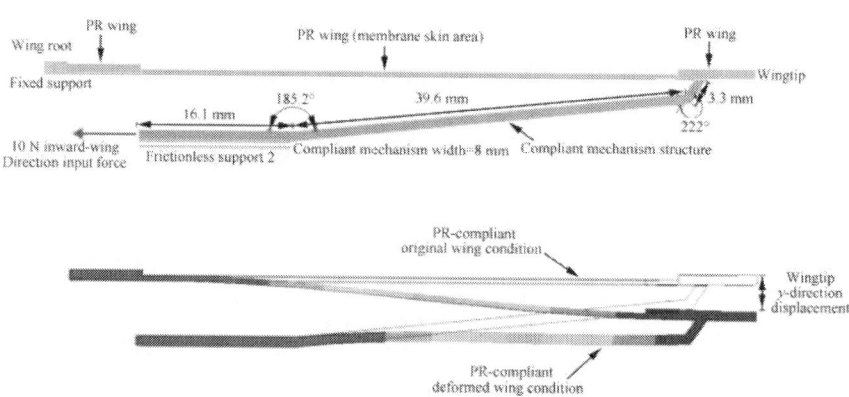

Figure 3. Compliant mechanism structure (top) and its kinematic principle (bottom) clipped at $x/c = 0.6$.

Generally, the PR wing (membrane wing configuration) and rigid wing (rigid wing configuration) used in the present works are almost identical to the PR-compliant wing, in terms of planform shape and dimension. The distinctive parts among the wings are the complaint mechanism structure and flexible membrane skin component. The PR wing model is designed by subtracting the compliant mechanism structure, whereas the rigid wing model is designed by removing the membrane skin and compliant mechanism component of PR-compliant wing. Summary of basic design dimension and configuration for all wings used in related works are given in Table 1.

Table 1. Basic design dimension and configuration for all MAV wing types.

Parameter	Rectangular wing (rigid wing)	PR-compliant (morphing wing)	PR (membrane wing)	Rigid (rigid wing)
Wingspan b (mm)	150	150	150	150
Root chord c (mm)	150	150	150	150
Aspect ratio A	1.25	1.25	1.25	1.25
Maximum camber at the root	Flat	6.7% of c (at $x/c = 0.3$)	6.7% of c (at $x/c = 0.3$)	6.7% of c (at $x/c = 0.3$)
Maximum reflex at the root	Flat	1.4% of c (at $x/c = 0.86$)	1.4% of c (at $x/c = 0.86$)	1.4% of c (at $x/c = 0.86$)
Built-in geometric twist (°)	0	0.6	0.6	0.6
Compliant mechanism component	Excluded	Included	Excluded	Excluded
Membrane skin component	Excluded	Included	Included	Excluded

The thickness for a rectangular wing model is set accordingly to experimental data at 2.94 mm whereas all other wing structures (including the compliant mechanism) and membrane skins are set at 1.0 mm and 0.5 mm, respectively. The origin of the coordinate system is located at the center of the leading edge of the wing and the following coordinate system is adopted: x is chordwise, z is spanwise, and y is normal to the wing.

Materials selection, boundary conditions and mesh generation for static structural analysis
Aluminum 1100-0 and rubber are adopted as the material for all wing structure (including the compliant mechanism component) and membrane skin, respectively. Selection of aluminum is based on its excellent forming characteristics, machinability, being highly resilient and its previous application on compliant mechanism components.[21] The material properties of aluminum 1100-0 and rubber are listed in Table 2. Instead of using a hyperelasticity material model, the rubber material is modeled as a linear elastic model for simplification.[22]

Table 2. Material properties of aluminum 1100-0 and rubber.

Material property	Aluminum 1100-0	Rubber
Density (kg/m³)	2707	1000
Young modulus (Pa)	6.9×10^9	8.642×10^6
Poisson's ratio	0.33	0.49
Bulk modulus (Pa)	6.76×10^9	1.44×10^8
Shear modulus (Pa)	2.59×10^9	2.90×10^6
Tensile Yield strength (Pa)	3.5×10^7	1.3787×10^7

In order to replicate the twist morphing condition on PR-compliant wing model, few displacement constraints are utilized and combined with 10 N inward-wing retraction force (B label), as illustrated in Fig. 4. The 10 N inward-wing retraction force is applied based on structural optimization study, in which, the amount of this force has sufficiently produced a significant wingtip displacement and elucidates the morphing wing condition. A fixed support is employed at the root chord area ($x/c = 0.33$–0.67) in order to simulate the restrictive effect of sting balance equipment attachment region. [6] This boundary condition is intentionally defined to reflect future works of real wind tunnel testing. The similar boundary conditions are applied to all other wings configuration, excluding the boundary condition components related to compliant mechanism structure such as the retraction force and frictionless support No. 2 (E label) in Fig. 4.

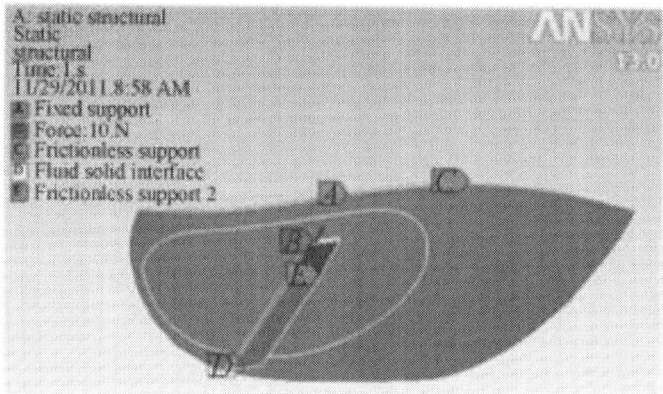

Figure 4. Static structural boundary conditions on PR-compliant wing.

Unstructured tetrahedral mesh with ANSYS SOLID 187 3D element type is created on all wing models. The grid independent study results an optimized grid around 80000 elements for static structural analysis as shown in Fig. 5.

Figure 5. Elements for static analysis created on PR-compliant wing.

Airflow boundary conditions and mesh generation

The computational flow domain is built around an MAV wing, in which the symmetrical condition is manipulated by modeling only half of the computational domain. The 3D boundary of the computational flow domain is dimensioned in the root chord length c and placed remotely away from the MAV surface to ensure no significant effect on the aerodynamics, as shown in Fig. 6. An initial model with 200000 unstructured elements is created and used to solve the airflow field. The grid independent test results show that the optimized grid is achieved at 1000000 elements as depicted in Fig. 7. The growing prism inflation layer option has been implemented on fluid–solid boundaries with the first cell above the wall set at $y^+ \leqslant 1$.

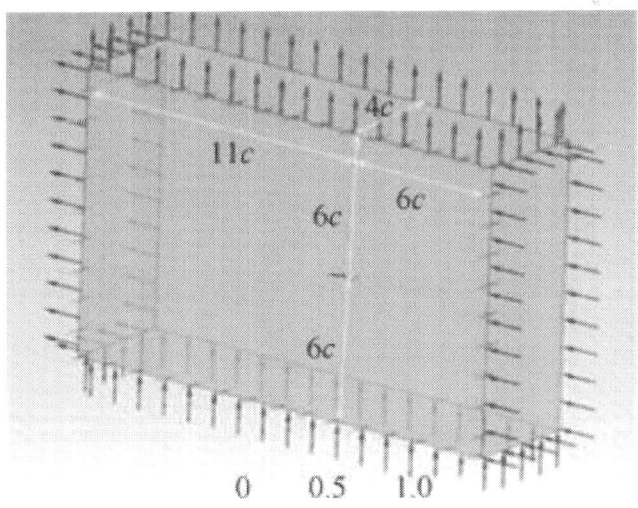

Figure 6. Computational flow domain.

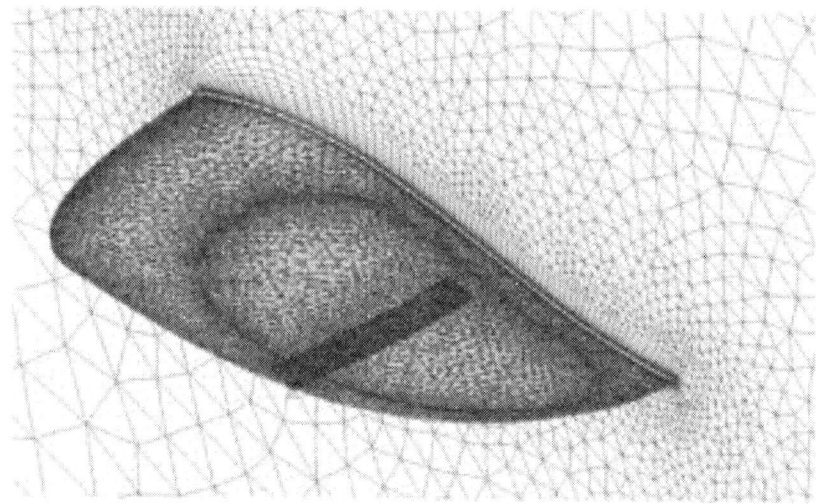

Figure 7. Elements for CFD analysis.

The inlet and outlet are marked by the flow vectors (see Fig. 6), with velocity magnitude of 9.696 m/s (which is equivalent to Reynolds number, $Re = 100000$ at chord, and common operational Re for MAV) is specified at the inlet, and zero pressure boundary condition is enforced at the outlet. The angle of attack (AOA) is varied between $-5°$ and $30°$ ($-7°$ to $40°$ for rectangular rigid wing verification case). The symmetrical wall and side walls are assigned as symmetrical boundary condition and slip surface boundary condition, respectively. The wing surface itself is modeled as a no-slip boundary surface and satisfies the quasi-static aeroelastic condition for FSI investigation. The turbulence intensity of 5% with automatic wall function is fully employed to solve the flow viscous effect.

RESULTS AND ANALYSIS

Rectangular wing verification results

The purpose of this verification is to justify and ascertain the method's validity in terms of pre-processing setup and boundary condition selections. The verification is accomplished by comparing the numerical results to a reliable experimental data from Mueller and Torres[18].

Since this rectangular wing is declared as a rigid wing configuration, one could expect that this wing should produce minimal amount of deformation. This expectation is justified through the wing plane deformation result shown in Fig. 8. Despite the deformation (displacement on y-direction) pattern has revealed a linear increase from wing root to the wing-tips, yet the maximum magnitude of this deformation is fairly small (below 0.0005% of the root chord). As a result, the details of wing deformation characteristics in terms of aerodynamic twist (maximum camber location and magnitude of maximum camber) and geometric aerodynamic twists for this wing are assumed to be minute, with insignificant influence on overall aerodynamic performance.

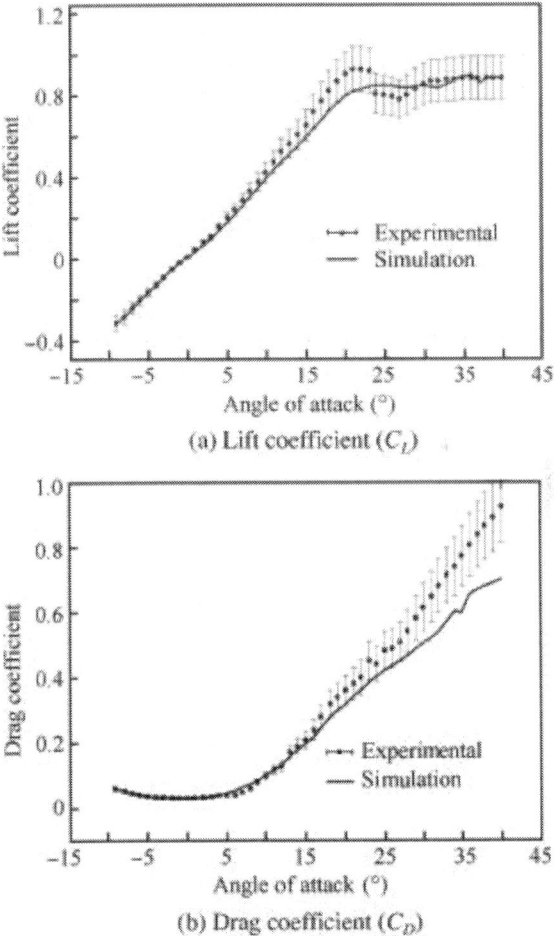

(a) Lift coefficient (C_L)

(b) Drag coefficient (C_D)

Figure 9. Verification results between the FSI numerical result and experimental data from Mueller and Torres[18].

A comparative study has been performed to evaluate the lift and drag coefficient performance between the simulation results against the experimental data, as shown inFig. 9. Fig. 9(a) points out that almost all lift coefficient values (at AOA −7° to 40°) for the rectangular wing simulation, lie within a defined 10% error margin, compared to experimental data. The numerical method has also performed well in predicting the stall angle for this low aspect ratio wing since both results matched to show a stall delay at relatively high angles (20–23°). Both simulation and experimental data also exhibited very low lift slopes (about 0.041/(°)), and this is a common characteristic for a low aspect ratio wing.[6]

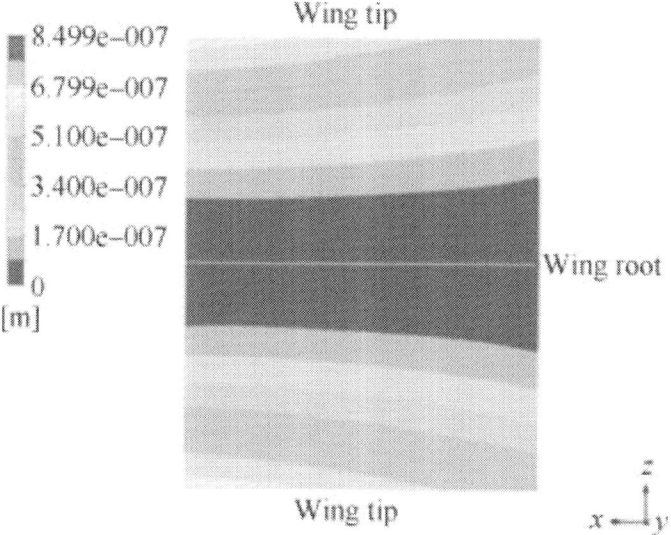

Figure 8. Out of plane deformation for rectangular wing.

The drag coefficient values were also sufficiently predicted in pre-stall angle ranges (i.e., −7° at 22°), although deviation due to under prediction is observed when AOA extends into post-stall angle (23–40°). This is due to RANS modeling and SST turbulent model used in present study. This phenomena are expected for steady RANS–SST formulation, in which, at very high AOA, the influence of organized transient motion could not be taken into account by this formulation.[23] and [24] These results provide evidence that the numerical model has shown good capability to produce satisfactory correlation results in predicting the lift coefficient and drag penalty over the rectangular wing. It is safe now to conclude that this numerical framework is verified to be superior enough to be employed in further related works.

PR-compliant wing design performance

Wing structural deformation characteristics

The plane deformations (y-direction displacement) for PR-compliant, PR and rigid wing at 0°, 10° and 20° are depicted in Fig. 10. Results clearly show the upward deflection on the membrane region (also known as membrane inflation) has increased for PR-complaint and PR wing model when the AOA increases. However, the maximum upward deflection produced by both wings is relatively small, i.e., at 3% of the root chord. This membrane inflation wing state is consistent with previous works by Albertani and co-workers,[4] Shyy and co-workers [25], [26] and [27] and Standford and co-workers. [6],[22] and [28] In those works, the inflation is suggested to be contributed by the aerodynamic loads during flight.

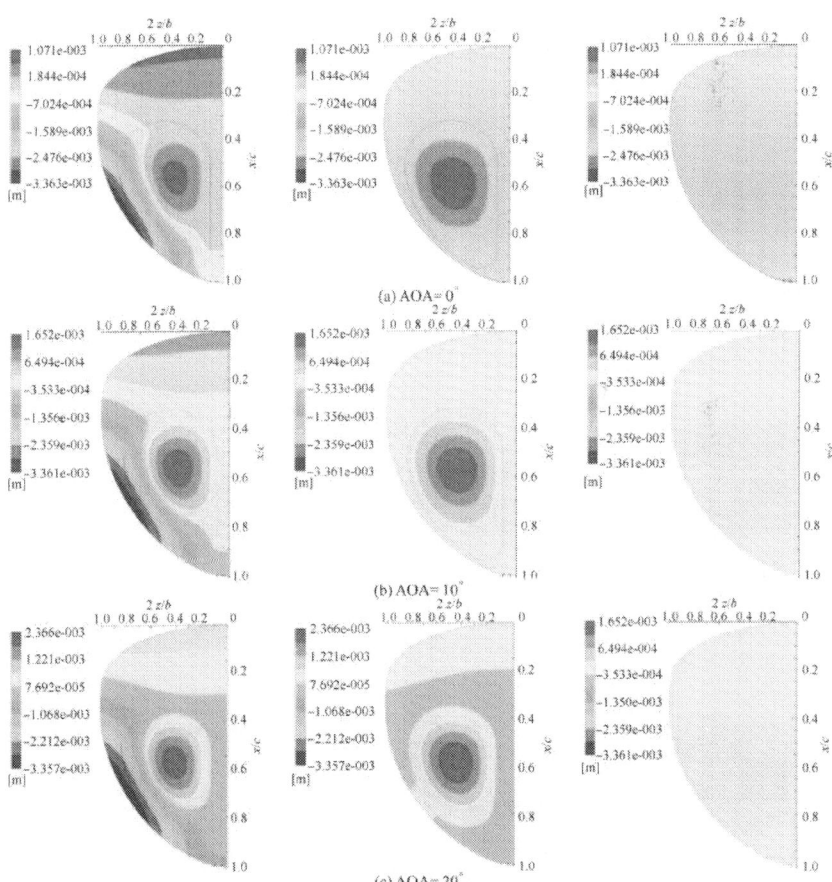

Figure 10. Plane deformation for PR-compliant (left), PR (middle) and rigid (right) wing.

It is also noticed that the deformation characteristic of rigid wing also results in an upward deflection, corresponding to AOA incidence. However, its maximum deflection magnitude is relatively small at 0.13% of root chord. Focusing on the PR-compliant wing's deformation the figure clearly shows that PR-compliant wing has produced a downward deformation at aft wing-tip between the region of $2z/b = 0.3–1.0$. The magnitude of this maximum downward deformation is about 3.4 mm (equivalent to 2.27% of root chord). This maximum downward deformation however, remains constant throughout the wing incidence variation. Therefore, the deformation is most likely contributed by the external force actuation rather than the aerodynamic loads.

In further understanding these wing deformation characteristics, a detailed study on the wing's aerodynamic twist performance (maximum camber location and magnitude of maximum camber) has been carried out. The chordwise cross-section of $2z/b = 0.47$ is preferred since it is concurrently positioned on maximum membrane inflation location and contributes highly to aerodynamic twist alteration. The results for the maximum camber location and the maximum camber magnitude for all wings are shown in Fig. 11 and Fig. 12. Both results apparently show that the performance of all wings (in this case) is similar throughout the AOA, with maximum deviation approximately at 3%. Based on these results, it can be construed that aerodynamic loading and external force (on PR-compliant wing configuration) has very little effect on the aerodynamic twist on all wings in this work.

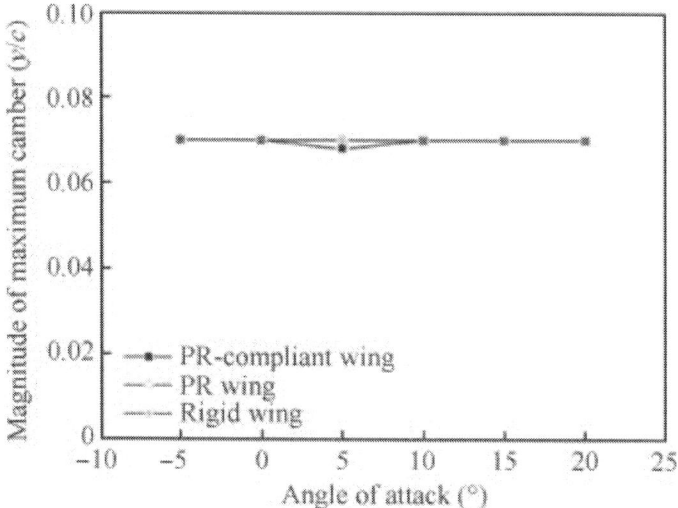

Figure 11. Maximum camber location characteristics for all wings taken at $2z/b = 0.47$.

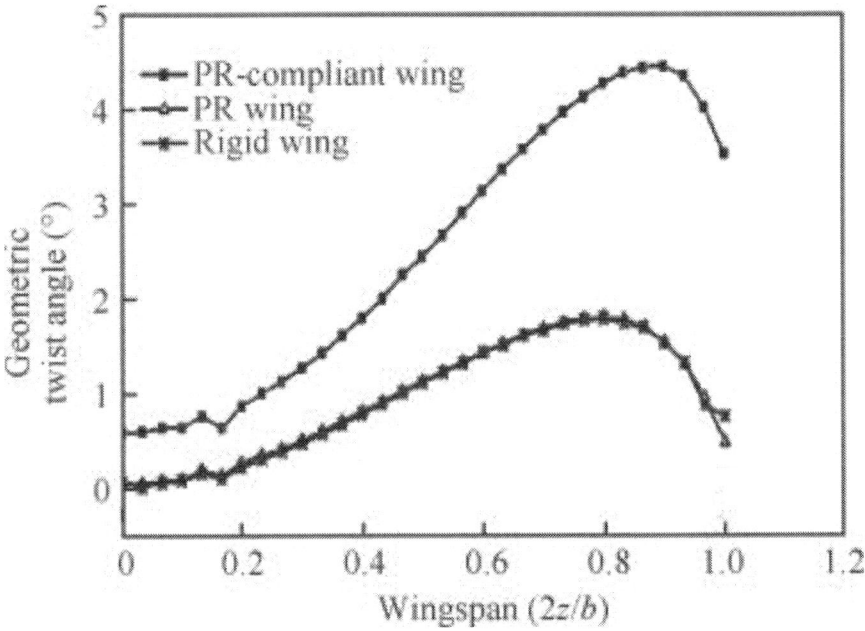

Figure 12. Magnitude of maximum camber characteristics for all wings taken at $2z/b = 0.47$.

Further investigation on geometric twist characteristics for all wings has also been carried out. Geometric twist is defined as the twist of a wing that has a geometrical local angle of attack changes at different spanwise positions. Fig. 13 depicts the geometric twist characteristics for all wings at AOA = 10°. The results show that PR-compliant wings have the most substantial geometric twist performance among the wings, in which, for any chordwise wing cross-section, the magnitude of the PR-compliant wing's geometric twist is always more than of other wings by 50%. The maximum geometric twist for PR-compliant wing occurs at $2z/b = 0.9$ with magnitude of 4.45°, which is 65% higher than the initial twist condition shown by rigid wing. This twist condition is a direct result from the PR-compliant deformation pattern shown in Fig. 10, in which, the deformation has significantly increased the local AOA at every wingspan cross-section towards the wingtip. In aerodynamic study, this wing condition is designated as "washin" wing. Investigation on geometric twist characteristic is extended in order to clarify the influence of aerodynamic loads on PR-compliant twist condition.

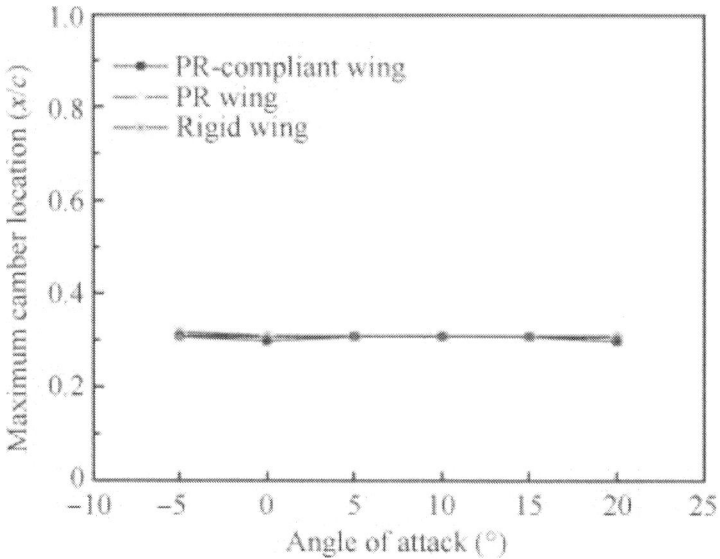

Figure 13. Geometric twist characteristics of all wings at AOA = 10°.

Fig. 14 represents the maximum geometric twist for all wings at different AOA. At this point, one can find that the maximum geometric twist characteristic for all wings is relatively a weak function of AOA changes. This situation indicates that the aerodynamic loads have minimal influence on the geometric twist characteristics for all wings.

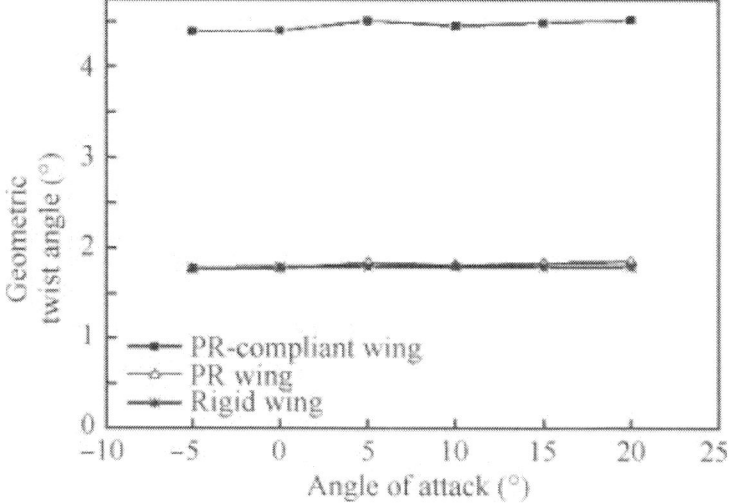

Figure 14. The maximum geometric twist for all wings at different AOA.

Based on this structural deformation, aerodynamic and geometric twist results, it was clearly suggested that the PR-compliant wing has the ability to create significant deformations, which in turn, contributes highly towing geometric twist behavior. Even though there are obvious signs of membrane inflation occurring on the membrane wing's area for PR-compliant and PR wings, respectively, the results does not demonstrate an enormous influence in modifying the aerodynamic twist or geometric twist characteristics on the wings. In understanding the role played by geometric twist alteration on PR-compliant wing, further investigation related to its aerodynamic performance is carried out in the following section.

PR-compliant wing aerodynamic characteristics
(1) Lift coefficient

Lift coefficient C_L distributions for all wings are depicted in Fig. 15. At this point, the lift coefficient for every wing has performed nonlinearly with AOA changes. This is common lift characteristic for low aspect ratio wings suggested by previous works, e.g. by Shields and Mohseni, [29] Mueller, [30] Pelletier et al., [31] Sathaye, [32] and Mueller and Torres [18].

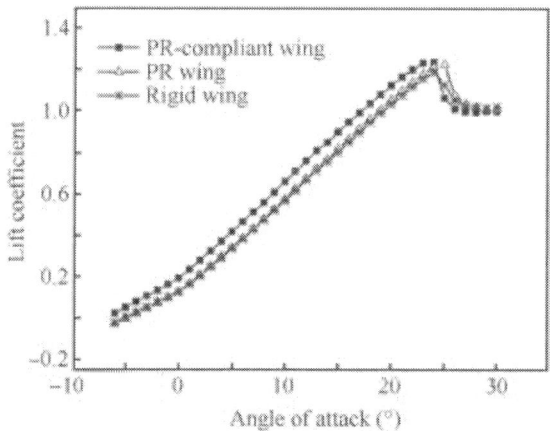

Figure 15. Lift performance for all wings.

In evaluating the nonlinear lift performance, PR-compliant wing has surprisingly produced better lift coefficient in every AOA up to stall angle. Analytically, the PR-compliant wing has an ability to generate about 20% more lift than to the other wings at AOA below 6°, but this percentage has monotonically decreased as the AOA increases. The PR-compliant wing manages to produce lift at the rate of 5%–19% more than other wings at the AOA region of 7° up to stall angle.

The FSI simulation has also managed to predict the stall angle for every wing at AOA region of 24–25°. The maximum lift coefficient for PR-compliant wing is marked at 1.236, which is about 2%–3% higher than all other wing types. In the early post-stall angle (1–2° after stall angle), all wings exhibited a sudden drop of lift coefficient distribution. The percentage of this drop is analyzed by taking the data points between the stall angle and the next angle point after stall. The analysis finds that PR-compliant wing has the most severe drop of lift coefficient at approximately 14% followed by PR and rigid wing at 12% and 5%, respectively.

(2) Drag coefficient
Total drag coefficients (C_D) for every wing at a sweep AOA = −6° to 30° are shown in Fig. 16. At this point, one can find that the drag coefficient for every wing is monotonically increased with AOA. At very low AOAs (between −5° to −1°), the drag coefficient distribution for all wings yields almost identical performance. As the AOA changes from 0° to stall angle however, the PR-compliant wing displays the highest drag penalty. PR-compliant's drag coefficient has increased, up to 29% more than the other wings at certain angles.

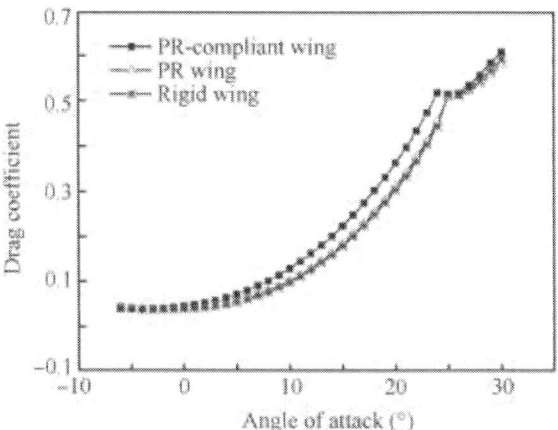

Figure 16. Drag distribution for all wings.

Analyzing the rate of drag increment for all wing types (starts from 0° to stall angle), it can be derived that for every 1° increment of AOA, an increase of approximately 5%–13% in drag coefficient is experienced. However, a halt of drag increment was seen at early post-stall angle (1–2° after stall). PR-compliant wing suffers most among the wings, with drop of drag coefficient at approximately 1%. Nonetheless, this sudden drop only occurs in early post-stall region before it returns to increase monotonically at the rate of 2%–4% at higher post-stall wing incidence.

(3) Moment coefficient
The longitudinal pitching moment characteristics (measured about the leading edge) for all the wings are given as a function of C_L as depicted in Fig. 17. As expected, the pitching moments for all wing have a negative slope, which indicates a nose-down moment that is essentially required for MAV longitudinal static stability performance. [6] Generally, one can find that the slope of the pitching moments for all the wings is a strong function of AOA. An analytical slope evaluation has been performed at two different stages. The first stage involves the slope analysis taken at data points from 0° to 15°(low-medium angle) while at the second stage, the slope is taken from 16° to 22° (high angle).

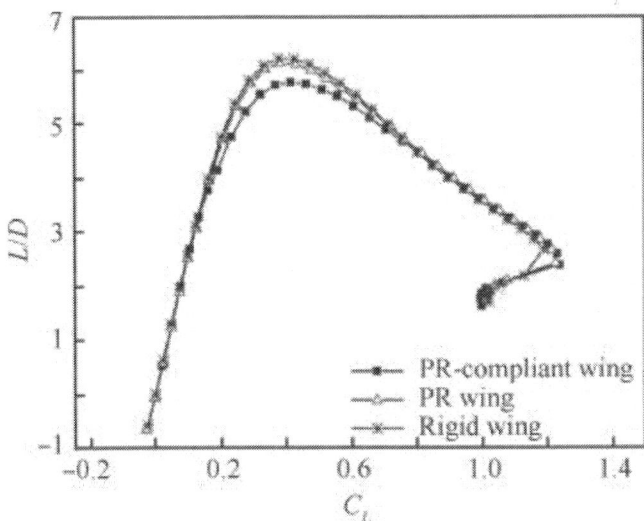

Figure 17. Longitudinal pitching moment characteristics for all wings.

At the first stage, both PR-compliant and PR wing performance are seen identical, with overlapping curve slopes of −0.261, respectively. This is approximately 5% steeper than rigid wing's slope. At the second stage, PR-compliant wing exhibits steeper slope level of −0.348, which is about 4.4% and 5.7% better than rigid and PR wings, respectively. The difference is relatively small; however the PR-compliant wing has shown a potential to produce more stability (through steeper pitching moment slope) compared to other wings.

(4) Lift to drag ratio

The lift to drag ratio (L/D) for all wings are given as a function of C_L in Fig. 18. In aerodynamic study, L/D is always used to signify the performance of wing aerodynamic efficiency. Fig. 18 shows that all the wings perform almost similarly at lift coefficient below 0.2 (equivalent to 3°). However, at above 0.2, the L/D curve for PR-compliant begins to deviate, giving lower L/D values, compared to PR or rigid wing curve. The peak efficiency for PR-compliant wing is achieved at AOA = 5° with $L/D = 5.771$, which is 6% lower than PR wing's peak efficiency.

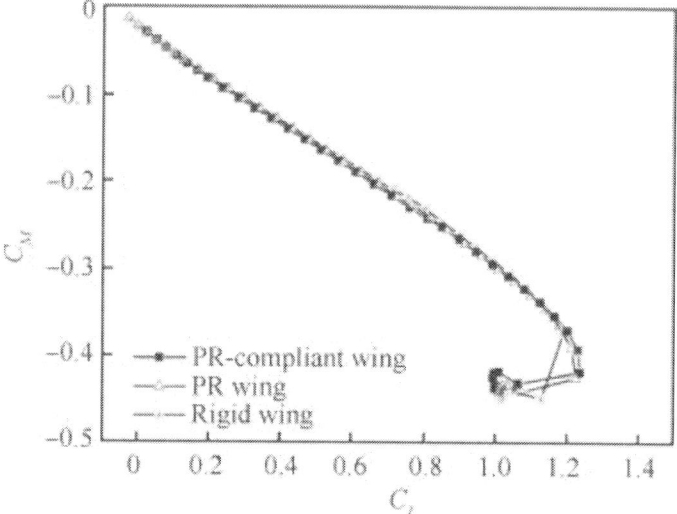

Figure 18. Lift to drag ratio distribution for all wings.

Based on this L/D performance, one can presume that PR-compliant wing has a lower aerodynamic efficiency compared to PR or rigid wing. According to earlier works by Shyyet al. [27] and Stanford et al. [28] they have suggested that the plunge of aerodynamic efficiency is most probably due to massive drag penalty created on membrane wing MAV, which has overwhelms the successive increase in lift generation.

Flow structures over PR-compliant wing

(1) Vortex formation

The vortices core regions over PR-compliant wing are visualized based on Q criterion (or the Q value), which is the second invariant of the velocity gradient tensor.[33] The vortices isosurface is shown in purple color to capture the regions of high swirl of $Q = 0.04$ at AOA = −5°, 0°, 5°, 10°, 20°, 25° as shown in Fig. 19. In the figure, TV means tip vortex.

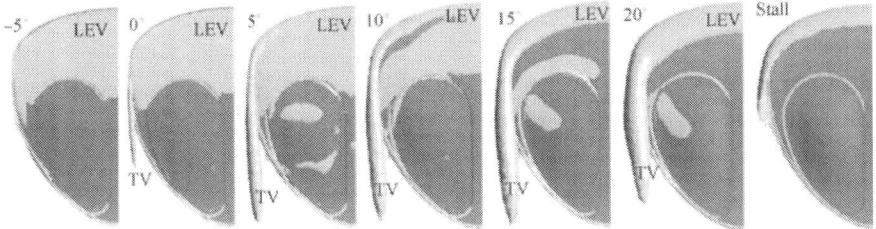

Figure 19. Vortices core region snapshot highlighted on PR-compliant wing's upper surface by the isosurface $Q = 0.04$ at $-5°$, $0°$, $5°$, $10°$, $20°$ and stall angle.

The results show that at very low AOA ($-5°$), PR-compliant wing manages to produce a dominant attachment of leading edge vortex (LEV) structure, which is seen present at almost half of the wing surface. However, no TV structure is spotted at this stage.

As the AOA increases to $0°$ from $-5°$, the LEV formation still shows a dominant attachment on the upper surface of the wing. At the same time, TV structure appears and can be clearly observed at the wing-tip area. At this point, LEV starts to show a weak interaction with TV, through a direct flow connection between them.

When the AOA increases further up to $20°$, TV structure is seen larger in diameter and extends at larger distances, towards the downstream aft of the wing-tips. Simultaneously, the LEV attachment on the wing surface is confined at the leading edge region, flowing directly into TV structure, which produces more interaction of LEV–TV formation.

At the early stall angle (AOA $= 25°$), result shows a severe vortex breakdown created on the wing surface. The earlier visibly TV structure at lower wing incidence is no longer observed and LEV structure is detached from the wing surface.

TV formation has an important effect on low aspect ratio wing, since naturally, it contributes to higher induced drag[27], whereas LEV structure has played an important role in providing an additional lift especially on small flapping animals such as bats[34] and swift.[35]

The interaction of LEV–TV generation over a low aspect ratio wing is well-known. This is very closely related to the nonlinear aerodynamic lift pattern. A strong downwash induced by a TV formation will force the

LEV structure to stay attached on the wing surface. This results in the presence of low pressure core region on the upper wing surface and consequently enhances the nonlinear lift formation.[33] and [36] These circumstances could be exemplified, apparently in the nonlinear lift coefficient pattern shown in Fig. 15 and the contour of pressure coefficient, C_p result as depicted in Fig. 20. The pressure coefficient result, C_p over all wings (Fig. 20, shown at AOA = 15°) has visibly demonstrated that the area of low pressure region on the PR-compliant wing has been increased, particularly at the wing-tip zone where the LEV–TV interaction takes place.

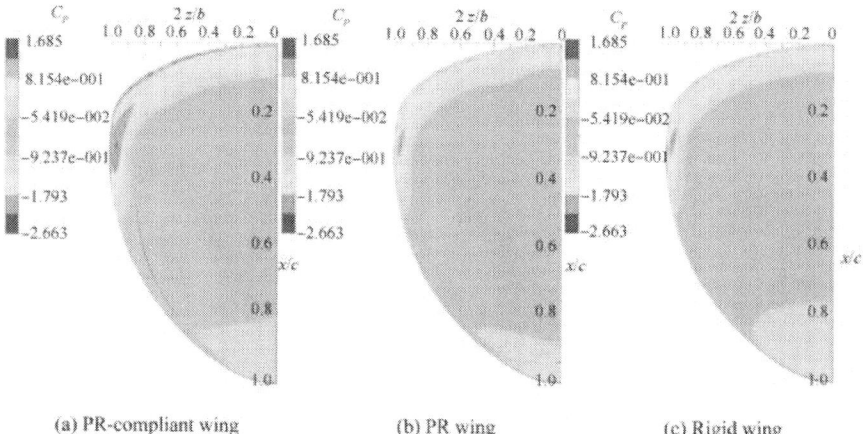

(a) PR-compliant wing   (b) PR wing   (c) Rigid wing

Figure 20. Contour of pressure coefficient (C_p) on upper wing surface at AOA = 15°.

The beginning of TV structure existence as early as AOA = 0° (Fig. 19) could also be viewed as the starting point of excessive drag penalty (Fig. 16) on PR-compliant wing. This situation is most probably due to strong association between TV formation and induced drag generation on low aspect ratio wing.[27] This phenomenon continues to intensify when the TV structure expands simultaneously according to the AOA changes. This results into higher total drag penalty contributed by the immense magnitude of induced drag.

The effect of vortex breakdown can be easily spotted in sudden reduction of lift (Fig. 15) and total drag (Fig. 16) coefficients at early post-stall angle. Deterioration of both LEV–TV interaction and TV formation has adverse effects on the generation of nonlinear lift and induced drag distribution at stall angle. The increase of total drag after the stall angle

(Fig. 16) is most probably associated with pressure drag component rather than induced drag. This situation strongly indicates that the nonlinear lift and drag distribution at high AOA are highly dependent on TV–LEV interaction and TV formation, respectively. Similar phenomenon can also be observed in the previous works done by Ringuette et al.,[37] Taira et al.[33] and Torres and Mueller.[38]

(2) TV strength and flow structure

Fig. 21 represents the TV low pressure intensity region taken at six sample planes (parallel to yz plane and located at 10 mm, 40 mm, 70 mm, 90 mm, 120 mm and 160 mm from leading edge) and upper wing flow streamline. This is done to elucidate the variation of tip vortex strength on all wings. The streamline is positioned in the vicinity of upper wing surface and flows from right to left direction. According to Shyy et al.,[27] the strength of TV formation is usually associated with the magnitude of low pressure core region. Thus, in this investigation, the strength of TV is evaluated based on the magnitude of low pressure core region, whereas the streamline flow is qualitatively examined in terms of flow attachment and streamwise flow deviation.

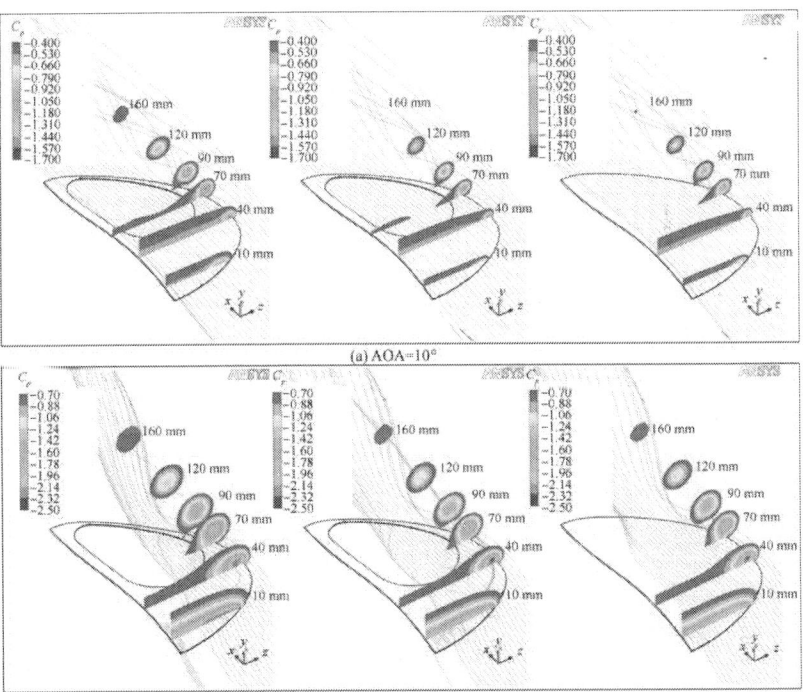

(a) AOA=10°

(b) AOA=20°

Figure 21. Low-pressure intensity in TV core region and streamline flow structure for PR-compliant (left), PR (middle) and rigid (right) wing at 10° and 20°.

At AOA = 10°, PR-compliant wing is able to generate the lowest pressure coefficient in the vortex core region with Cp = −1.7. This magnitude is approximately 44%, lower than other wings' pressure core and can be visibly seen at PR-compliant wing's 40 mm and 70 mm sample plane. Additionally, only PR-compliant wings are observed to delay the low pressure core downstream, apparently at 160 mm sample plane.

At 20° incidence angle, the lowest pressure coefficient for PR-compliant wing is approximately at Cp = −2.5 but this pressure core magnitude has a least difference compared to the other wings. Therefore, the strength of this vortex core region can only be examined qualitatively through streamline flow deviation behavior.

In investigating the streamline flow structure behavior, the results show that at AOA = 10°, the streamwise flows for all the wings are observed to be steadily attached on the wing surface. Even though the PR-compliant wing has the lowest pressure core magnitude among the wings, it has an almost identical streamline flow behavior compared to other wings, in which, only minimal circulation and least streamwise flow deviation are performed.

At AOA = 20°, the PR-compliant wing has produced a higher degree of flow circulation, which can be perceived on the wing's leading edge area and the wing-tip. This circulation has interacted with strong swirling flow at the wing-tip, which has induced the chordwise flow deviation that also known as spanwise flow (crossflow). This circumstance has signify that a higher TV strength is produced by PR-compliant wing, because a higher TV strength would encourage more spanwise or crossflow on the wing surface.[29] and [33].

CONCLUSIONS

(1) A two way FSI investigation consists of quasi-static aeroelastic structural analysis couple with 3D incompressible RANS–SST solver is used to solve the wing aerodynamics for steady, incompressible flow over a low Reynolds number and low aspect ratio rectangular, PR-compliant, PR and rigid wing.

(2) Verification of numerical method on a rigid rectangular wing achieves a good correlation with available experimental results in terms of lift distribution, drag penalty and overall structural deformation.

(3) Structural deformation results on PR-compliant wing shows a significant deformation, which has drastically contributes to a considerable wing geometric twist behavior compared to the other wing types. The membrane inflation due to aerodynamic loading on all membrane wings (PR-compliant and PR wing) has unable to demonstrate substantial influence in modifying the aerodynamic twist or geometric twist characteristics on the wings.

(4) Aerodynamic investigation on PR-compliant wing has shown its ability in generating a higher nonlinear lift distribution compared to PR or rigid wing type. This benevolent performance is most probably due to strong LEV–TV interaction on PR-compliant, which has induced the low pressure core region on the upper wing surface.

(5) Despite the superior lift performance, PR-compliant wing also suffers from larger drag penalty than the other wings. This situation has a close connection to the strong TV formation on PR-compliant wing, in which intense TV formation promotes higher induced drag generation. Consequently the overall aerodynamic efficiency for PR-compliant wing is plunged due to substantial drag penalty, which has overwhelms the successive increase in lift generation.

(6) Based on the vortices structure and streamline flow investigation, the numerical results show that PR-compliant wing has a greater level of TV strength, in which has encouraged more spanwise flow on the wing surface. The vortices deterioration at stall angle has abruptly plunged the generation of nonlinear lift and drag distribution. This situation has strongly indicates that the lift and drag distributions on low aspect ratio wing enormously depend on vortices formation especially at high AOA.

Future work will include the variation of force for morphing activation with experimental validation of the aeroelastic wing deformation and flow structures. Additionally, the morphing wing structures incorporating the geometric twist will be physically examined.

ACKNOWLEDGEMENTS

The authors acknowledge financial support from the Government of Malaysia via the sponsorship by the Ministry of Higher Education under the IPTA Academic Training Scheme awarded to the first author. The financial support provided by the Malaysia Ministry of Higher Education's Fundamental Research Grant Scheme (FRGS) (No. 600-RMI/FRGS 5/3 (22/2012)) is also acknowledged.

REFERENCES

1. Mohan S, Sridharan G. Emerging technologies for microunmanned air vehicles. Defence Sci J 2001;51(3):223–8.

2. Viieru D, Albertani R, Shyy W, Ifju P. Effect of tip vortex on wing aerodynamics of micro aerial vehicles. AIAA J 2005;42(4):1530–6.

3. Lian Y, Shyy W. Laminar–turbulent transition of a low Reynolds number rigid or flexible airfoil. AIAA J 2007;45(7):1501–3.

4. Albertani R, Stanford BK, Hubner JP, Ifju PG. Aerodynamic coefficients and deformation measurements on flexible micro air vehicle wings. Exp Mech 2007;47(5), 625–5.

5. Stanford BK. Aeroelastic analysis and optimization of membrane micro air vehicle wings [dissertation]. Gainesville: University of Florida; 2008.

6. Stanford BK, Ifju P, Albertani R, Shyy W. Fixed membrane wings for micro air vehicles: experimental characterization, numerical modeling, and tailoring. Prog Aerosp Sci 2008;44(4), 258–4..

7. Abdulrahim M, Garcia H, Lind R. Flight characteristics of shaping the membrane wing of a micro air vehicle. J Aircraft 2005;42(1):131–7.

8. Song A, Tian X, Israeli E, Galvao R, Bishop K, Swartz S. Aeromechanics of membrane wings with implications for animal flight. AIAA J 2008;46(8):2096.

9. Supekar AH. Design, analysis and development of a morphable wing structure for unmanned aerial vehicle performance augmentation [dissertation]. Arlington: University of Texas; 2007.

10. Combes SA, Daniel TL. Flexural stiffness in insect wings II. Spatial distribution and dynamic wing bending. J Exp Biol 2003;206(17): 2989–7.

11. Wlezien RW, Horner GC, McGowan A-MR, Padula SL, Scott RJ, Silcox RJ. The aircraft morphing program. AIAA J 1998.

12. Sofla AYN, Meguid SA, Tan KT, Yeo WK. Shape morphing of aircraft wing: status and challenges. Mater. Des 2010;31(3):1284–92.

13. Weisshaar TA. Morphing aircraft technology – new shapes for aircraft design. In: Proceedings of multifunctional structures/ integration of sensors and antennas; 2006.

14. Mujahid A. Dynamic characteristics of morphing micro air vehicles [dissertation]. Gainesville: University of Florida; 2004.

15. Shuib S, Ridzwan MI, Kadarman AH. Methodology of compliant mechanisms and its current developments in applications: a review. Am J Appl Sci 2007;4(3):160–7.

16. Abdulrahim M, Garcia H, Ivey GF, Lind R. Flight testing a micro air vehicle using morphing for aeroservoelastic control. In: Proceeding of 45th AIAA/ASME/ASCE/AHS structures, structural dynamics, and materials conference; 2004.

17. Anon. ANSYS CFX-solver theory guide. Canonsburg, PA: ANSYS, Inc.; 2010; 724-46.

18. Mueller TJ, Torres GE. Aerodynamics of low aspect ratio wings at low Reynolds numbers with applications to micro air vehicle design and optimization. Hessert Center for Aerospace Research, Department of Aerospace and Mechanical Engineering; 2001. Report No.: UNDAS-FR-2025. Contract No.: N00173-98-C-2025.

19. Albertani R. Experimental aerodynamic and static elastic deformation characterization of low aspect ratio flexible fixed wings applied to micro aerial vehicles [dissertation]. Gainesville: University of Florida; 2005.

20. Palmisano J, Ramamurti R, Lu K-J, Cohen J, Sandberg W, Ratna B. Design of a biomimetic controlled-curvature robotic pectoral fin. In: Proceeding of IEEE international conference on robotics and automation; 2007.

21. Kota S, Osborn R, Ervin G, Maric D, Flick P, Paul D. Mission adaptive compliant wing – design, fabrication and flight test mission adaptive compliant wing. In: Proceeding of RTO applied vehicle technology panel (AVT) symposium, evora, portugal; 2009.

22. Stanford BK, Albertani R, Ifju P. Static finite element validation of a flexible micro air vehicle. Experimental Mechanics 2007;47(2):283–94.

23. Benim A, Pasqualotto E, Suh SH. Modelling turbulent flow past a circular cylinder by RANS, URANS, LES and DES. Prog. Comput. Fluid Dyn Int J 2008;8(5):299–307.

24. Benim AC, Cagan M, Nahavandi A, Pasqualotto E. RANS predictions of turbulent flow past a circular cylinder over the critical regime. In: Proceeding of 5th IASME/WSEAS international conference on fluid mechanics and aerodynamics; 2007.

25. Shyy W, Aono H, Chimakurthi SKK, Trizila P, Kang C-K, Cesnik CESES. Recent progress in flapping wing aerodynamics and aeroelasticity. Prog Aerosp Sci 2010;46(7):284–327.

26. Shyy W, Lian Y, Chimakurthi SK, Tang J, Cesnik CES, Stanford B, et al. Flying insects and robots. Berlin, Heidelberg: Springer, Berlin Heidelberg; 2009. p. 143–57.

27. Shyy W, Ifju P, Viieru D. Membrane wing-based micro air vehicles. Appl Mech Rev 2005;58:283–301.

28. Stanford BK, Ifju P. Membrane micro air vehicles with adaptive aerodynamic twist : numerical modeling. J Aerospace Eng 2009;22(2):173–84.

29. Shields M, Mohseni K. Effects of sideslip on the aerodynamics of low-aspect-ratio. AIAA J 2012;50(1):85–99.

30. Mueller TJ. Aerodynamic measurements at low Reynolds numbers for fixed wing micro-air vehicles. Hessert Center for Aerospace Research, Department of Aerospace and Mechanical Engineering; 1991. Contract No.:N00173-98-C-2025.

31. Pelletier A, Mueller TJ, Dame N. Low Reynolds number aerodynamics of low-aspect-ratio, thin/flat/cambered-plate wings. J Aircraft 2000;37(5):825–32.

32. Sathaye SS. Lift distributions on low aspect ratio wings at low Reynolds numbers [dissertation]. Worcester: Worcester Polytechnic Institute; 2004.

33. Taira K, Colonius T. Effect of tip vortices in low-Reynoldsnumber poststall flow control. AIAA J 2009;47(3):749–56.

34. Muijres FT, Johansson LC, Barfield R, Wolf M, Spedding GR, Hedenstro¨m A. Leading-edge vortex improves lift in slow-flying bats. Am Assoc Adv Sci 2008;319(5867):1250–3.

35. Videler JJ, Stamhuis EJ, Povel GDE. Leading-edge vortex lifts swifts. Science 2004;306(5703):1960–2.

36. Taira K, Colonius T. Three-dimensional flows around low-aspectratio flat-plate wings at low Reynolds numbers. J. Fluid Mech. 2009;623:187–207.

37. Ringuette MJ, Milano M, Gharib M. Role of the tip vortex in the force generation of low-aspect-ratio normal flat plates. J Fluid Mech 2007;581(1):453–68.

38. Torres G, Mueller T. Low aspect ratio wing aerodynamics at low Reynolds numbers. AIAA J 2004;42(5):865–73.

CITATION

N.I. Ismail, A.H. Zulkifli, M.Z. Abdullah, M. Hisyam Basri, Norazharuddin Shah Abdullah, Computational aerodynamic analysis on perimeter reinforced (PR)-compliant wing, Chinese Journal of Aeronautics, Volume 26, Issue 5, October 2013, Pages 1093-1105, ISSN 1000-9361, http://dx.doi.org/10.1016/j.cja.2013.09.001.

CHAPTER 5

A Spongy Icing Model for Aircraft Icing

Xin Li[1], Junqiang Bai[1], Jun Hua[2], Kun Wang[1], Yang Zhang[1]

[1] School of Aerodynamics, Northwestern Polytechnical University, Xi'an 710072, China
[2] Chinese Aeronautical Establishment, Beijing 100012, China

ABSTRACT

Researches have indicated that impinging droplets can be entrapped as liquid in the ice matrix and the temperature of accreting ice surface is below the freezing point. When liquid entrapment by ice matrix happens, this kind of ice is called spongy ice. A new spongy icing model for the ice accretion problem on airfoil or aircraft has been developed to account for entrapped liquid within accreted ice and to improve the determination of the surface temperature when entering clouds with supercooled droplets. Different with conventional icing model, this model identifies icing conditions in four regimes: rime, spongy without water film, spongy with water film and glaze. By using the Eulerian method based on two-phase flow theory, the impinging droplet flow was investigated numerically. The accuracy of the Eulerian method for computing the water collection efficiency was assessed, and icing shapes and surface temperature distributions predicted with this spongy icing model agree with experimental results well.

INTRODUCTION

Ice accretion is a common and an important feature in flight. The presence of ice accretion can cause a serious safety problem; the most severe penalties encountered deal with increased drag, decreased lift, decreased stall angle, and reduced controllability. On the safety issues, aircraft icing has caused more than 50 accidents in the US in recent 20 years. The ice accretion problems can be studied by the wind tunnel testing and the real flight test, or the engineering and numerical approaches. The real flight test and the wind tunnel testing require extensive analyses, which are expensive and dangerous. The engineering approach employs empirical formulation and experimental data, which is much simpler but lack of precision. Therefore, the numerical method is widely adopted for its economical, efficient, and accurate features.

Conventionally, the simulation of ice accretion is based on the Lagrangian particle-tracking technique for the trajectory calculation and employing Messinger icing model. Bougault et al. developed an Eulerian method for the ice accretion.[1] After that, Eulerian method became popular because of its simplicity and efficiency. However, better numerical methods are needed to track and collect droplets and further studies are needed to understand the key factors on the icing growth. Currently, a number of ice accretion codes have been well developed by some international icing communities, such as ONERA (France), CIRAML (Italy), DRA (United Kingdom), LEWICE (USA), and FENSAP-ICE (Canada). The FENSAP-ICE code employs the Eulerian method, and the others are using the Lagrangian method.

The ice accretion problem has been studied widely in the last several decades, and many work had been done on prediction of ice shape by many researchers such as Bragg, Shen, and Shin and Bond.[2,3 and 4] In 1985, Bragg made some improvements on his previous model,[5] derived a method to solve the droplet trajectories and gave some recommendations for further improvement of the method. LEWICE was developed by the University of Dayton in 1983. Potapczuk and Bidwell extended the original LEWICE based on potential flow analysis to LEWICE/NS which used the solution of 2D Navier–Stokes equations.[6] Besides, it has been implemented to solve the energy equation to obtain the heat-transfer coefficient automatically. Another icing code from DRA has also been developed, especially for the ice protection on airfoils and on rotor blades of a helicopter. Recently, an ALE mesh movement scheme for long-term in-flight ice accretion has been performed by Fossati et al.[7]

The Lagrangian approach provides a good result but there are still some shortages. For example, it is hard to be applied on some complicated geometries, such as 3-D aircraft wing and multi-element airfoils. Bourgault developed an Eulerian approach on ice accretion[1] to overcome the shortages. On modeling the aircraft icing, the classical one was developed by Messinger.[8] Then Myers made some improvements on Messinger model.[9 and 10] Also, Bourgault and Otta et al. developed icing model for aircraft respectively.[11 and 12] With regard to the existing codes, FENSAP-ICE employs the icing model developed by Bourgault,

ICECREMO employs Myers's lubrication type model, and the icing model of LEWICE is based on Messinger's work.

The icing model developed in this paper included two physical phenomena that traditional icing model ignores: one is the supercooling of accreting ice surfaces, and the other is the sponginess of ice. Experiments have already confirmed that the accreting surface temperature would fall below the freezing point when ice started to form.[13] Karev et al. investigated the icing growth on a cylinder by the wind tunnel testing,[14] and the surface temperature was found below the freezing point of water. The possibility of liquid entrapment by an ice accretion's growing matrix has recognized by List.[13] When droplets were entrapped in the ice matrix, this type of icing is named as the spongy icing. To account for the sponginess and the supercooling, Blackmore and Lozowski[15] proposed a theoretical spongy spray icing model with surficial structure in the field of atmospheric icing. The supercooling of the water film and sponginess of ice affect the ice accretion process, and determine the ice growth rate and the ice shape, so the two physical phenomena which are usually ignored by traditional icing models must be considered in aircraft icing. Besides, like many other icing model, the spongy icing model is based on continuity equation and heat balance equation, so it has good succession.

DESCRIPTION OF THE PROBLEM

To verify and validate the Eulerian droplet tracking method and the collection efficiency, a suitable test case must be considered. A comparison between the Eulerian method and LEWICE is presented. In addition, a comparison with experimental impingement data from Papadakis et al.[16] is discussed as well. The impingement data is presented in the form of LWC distribution and water collection efficiency. The experiments were performed with different MVDs for a NACA23012 airfoil and an iced NACA23012 airfoil. The test conditions are shown in Table 1. For the NACA23012 airfoil, the selected MVDs are 20 and 111 μm, while for the iced NACA23012 airfoil, the MVDs are 20, 52, and 111 μm. Moreover, the Eulerian approach and the icing model were validated by simulating icing conditions on a NACA0012 airfoil. The test temperatures are −4.4, −10, −13.3, and −19.4 °C, which cover both dry and wet regimes; details of the test conditions are given in Table 2. The predicted ice shapes were compared to the experimental results under the same conditions in the NASA Lewis Icing Research Tunnel and the numerical results from LEWICE.[4] At last, the prediction accuracy of the surface temperature was evaluated through the simulation on the surface of a non-rotating horizontal cylinder, which is 0.038 m in diameter. The computational results were compared with the experimental results in Ref.[14]; the test conditions for cylinder are shown in Table 3. Two duration times of ice accretion are set 15 and 15.4 min. The 15 min computation is used to compare ice shape with experiment, while 15.4 min computation is used to validate the surface temperature with experiment. In tables, AOA,V_∞, LWC,p_∞ and MVD are angle of attack, free-stream velocity, liquid water content, free-stream pressure and mean volumetric diameter respectively.

Table 1. Test conditions for droplets impingement.

Parameter	Value
AOA ($°$)	2.5
V_∞ (m/s)	78.22
LWC (g/m^3)	0.5
Chord (m)	0.9144
p_∞ (kPa)	94.8

Table 2. Test conditions for ice accretion on NACA0012.

Parameter	Value
Accretion time (s)	360
AOA ($°$)	4
MVD (μm)	20
V_∞ (m/s)	67.05
LWC (g/m^3)	1
Chord (m)	0.5334
p_∞ (kPa)	101.3

Table 3. Test conditions for ice accretion on cylinder.

Parameter	Value
T ($°$C)	17
MVD (μm)	43.4
V_∞ (m/s)	30
LWC (g/m^3)	3
p_∞ (kPa)	101.3

FORMULATION AND NUMERICAL METHOD

Based on the multiphase flow theory,[17] an Eulerian method to numerically simulate ice accretions is presented in this paper. Some assumptions are established as follows[1]:

(1) Droplets are spherical and complete without any breakage or deformation.
(2) No droplet collision/coalescence.
(3) No heat and mass transfer between the air phase and droplet phase.
(4) No turbulence effect on droplets.
(5) Gravity, air drag, and buoyancy are considered.

Governing equations for air phase

The flow field for air can be obtained by solving the Reynolds-Averaged Navier–Stokes equations. Numerical approach is based on the finite volume form of the integral equations. In a domain of a volume Ω with boundary Ω_a, let ρ, u, v, E, H and p be the density, Cartesian velocity components, total energy, total enthalpy, and pressure, the equation can be written in the integral form:

$$\frac{\partial}{\partial t} + \int\int_{\Omega} Q dV + \int_{\partial\Omega} F_c \cdot n dS = \frac{1}{R_e} \int_{\partial\Omega} F_v \cdot n dS \tag{1}$$

The vector of the conserved variables, the convective flux term, and the viscous flux term are given as follows

$$Q = \begin{bmatrix} \rho \\ \rho u \\ \rho v \\ \rho E \end{bmatrix}$$

$$F_c = \begin{bmatrix} \rho u & \rho v \\ \rho u^2 + P & \rho uv \\ \rho uv & \rho v^2 + P \\ \rho uH & \rho vH \end{bmatrix}$$

$$F_v = \begin{bmatrix} 0 & 0 \\ \tau_{xx} & \tau_{xy} \\ \tau_{yx} & \tau_{yy} \\ \Pi_x & \Pi_y \end{bmatrix}$$

$$\text{with} \begin{cases} \Pi_x = u\tau_{xx} + v\tau_{xy} - q_x \\ \Pi_y = u\tau_{yx} + v\tau_{yy} - q_y \end{cases}$$

Where τ_{xx}, τ_{xy}, τ_{yx}, τ_{yy} are the elements of the shear-stress tensor, q_x and q_y the elements of heat-flux vector.

The cell-centered finite volume method is employed to solve the Navier–Stokes. The lower–upper symmetric Gauss–Siedel (LU-SGS) algorithm for time marching and Roe scheme for the spatial discretization of the convective flux were implemented in codes. The treatment of the far field boundary condition is based on the introduction of Riemann invariants for a one-dimensional flow normal to the boundary. Isothermal wall boundary condition is applied. Chi et al.[18] suggested that Spalart–Allmaras was proper for simulating ice accretion in the air. The Navier–Stokes equations were closed by the S–A model with thermally perfect gas and temperature-dependent properties.

Governing equations for droplets

The droplets distributed in the flow field can be regarded as a kind of pseudo fluid which penetrates in the air flow field. The governing equations for droplets can be written as follows

$$\frac{\partial \rho_d \varphi}{\partial t} + \mathrm{div}(\rho_d V_d \varphi) = S_\varphi \tag{2}$$

Where $\varphi = \begin{bmatrix} 1 & v_{dx} & v_{dy} \end{bmatrix}$ denoting the variables in the droplet continuity equation and momentum equation respectively, ρ_d the droplet apparent density (the mass of droplets per unit volume) and $\rho_d = a_v \rho_w$, where a_v and ρ_w denote the droplet volume fraction and the density of water respectively, V_d denotes the droplet velocity vector and S_φ the source term.

The water collection efficiency on the surface of the leading edge is solved by the velocity distribution in the air flow field of the airfoil. The finite volume method is applied to discretizing the governing equations, the convective term is discretized using the quadratic upwind interpolation of convective kinematics (QUICK) scheme, and the deferred correction method is used to ensure the diagonal dominance in the discretized equations. The temporal term is discretized using the implicit scheme. The alternating direction implicit (ADI) iteration method is utilized to solve the algebraic equations. In order to guarantee the stability of the iterative solution of algebraic equations, the source term is disposed linearly.

A permeable wall boundary condition is applied to simulating the droplets impingement onto the wall. The far field boundary conditions for the flow field are introduced as follows:$\rho_d = LWC$, $v_{dx} = u_\infty$, $v_{dy} = 0$, where LWC represents the liquid water content.

On meshing, a single-zone C-type grid with wake cut is developed for the NACA23012 airfoil, the iced NACA23012 airfoil, and the NACA0012 airfoil; the grid has 601×177 grid points (see Figs. 1(a)–(c)) and extends 20 chord lengths away from the airfoil in each direction. A distance to the first grid point off the airfoil surface of 4×10^{-6} m is used to give a y^+ value of 0.5 approximately. A single-zone O-type grid is developed for cylinder; the grid has 601×165 grid points (see Fig. 1(d)) and extends 20 chord lengths away from the wall in all directions. A distance to the first grid point off the surface of 4×10^{-6} m is used to obtain the y^+ value between 0.1 and 0.8.

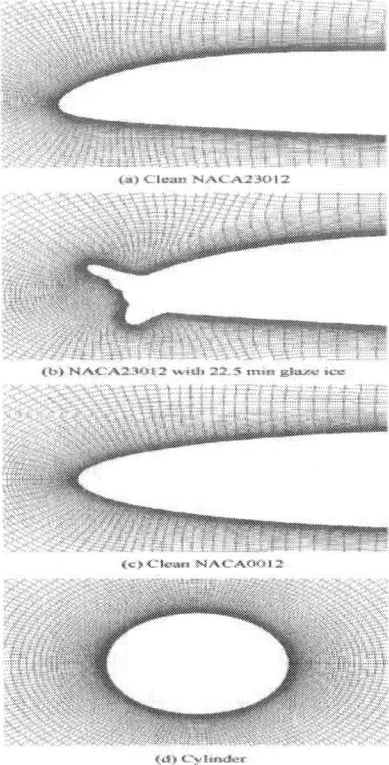

(a) Clean NACA23012

(b) NACA23012 with 22.5 min glaze ice

(c) Clean NACA0012

(d) Cylinder

Figure 1. Grid used in the present study.

CONCEPT AND STRUCTURE OF ICING MODEL

The following assumptions are considered; further experimental verification is needed to validate these assumptions.

(1) Each dendrite has a hemispherical tip and a cylindrical body.
(2) Ice accretes like dendrite crystals from the surface[19] (see Fig. 2).

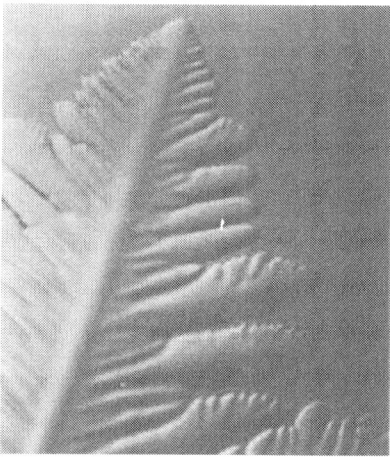

Figure 2. Microcosmic photograph of dendrite crystal.

(3) The radius of the hemispherical tip of the dendrite is proportional to the height of ice layer.[20]
(4) The latent heat of solidification produced by the ice forming layer is rejected to the water film layer.
(5) The radius of the dendrite is determined by the ambient temperature and the LWC.[19] However, when the LWC is low, the ambient temperature dominates.

Modeling of laminar layer film

The continuity equation for each control volume can be expressed as follows:

$$R_{in}+R_{imp}=R_{wf\text{-}ifl}+R_{out} \qquad (3)$$

where R_{in} is the total mass flux entering the control volume from the previous control volume, R_{imp} the flux of impinging droplets, and R_{out} the total mass flux flow out of the control volume. The mass flux of water transferred from the laminar layer to ice forming zone R_{wf-ifl}, equals the ice accretion flux $R_{wf-ifl} = I_0 + R_0$. The mass flux entrapped by the ice matrix from the ice forming zone R_{ifl-si} equals R_{wf-ifl}.

Dukler and Bergelin used the universal velocity distribution equation for turbulent flow of Von Karman in which the non-dimensional film flow rate was expressed by[21]:

$$u^+ = u_f/u_*$$ (4)

where u_f is the velocity of water films, and u_* the friction velocity. The non-dimensional coordinate normal to the substrate is

$$\zeta = \frac{\sqrt{y_t^3 g \rho_w}}{\mu_w}$$ (5)

where y_t is the water film thickness.

The Prandtl mixing length hypothesis results in several equations that are used to describe the velocity profile with the non-dimensional variable given in Eq. (5). The laminar layer is expressed by

$$u^+ = \zeta \quad (0 < \zeta \le 5)$$ (6)

The total mass flux in the film can be written as

$$R_t = \mu_w f(\zeta) L_{en}^{-1}$$ (7)

where L_{en} is the length of the control volume, R_t the total mass flux in the film

$$f(\zeta) = 0.5\zeta^2 \quad (0 \le \zeta \le 5)$$ (8)

For aircraft icing, $\zeta < 5$, which means the water film is laminar. In order to predict the pure ice growth rate and the liquid entrapment within the ice matrix, more details are needed to examine the ice forming layer. The assumptions in this section are applied. Further experimental verifications are needed to validate these assumptions.

Modeling of ice forming layer

Dendritic growth at the ice surface leads to a spongy ice. In List's icing model for a growing hailstone, there is a layer of ice lying between the spongy ice matrix and the surficial liquid. Here, we name this as an ice forming layer, which is assumed to contain some ice crystals that are growing with a uniform tip growth rate. Ice forming layer consists of water and dendritic ice, the thickness of the water film is $y_2 - y_1$ shown in Fig. 3, and all the temperatures are in degrees Celsius.

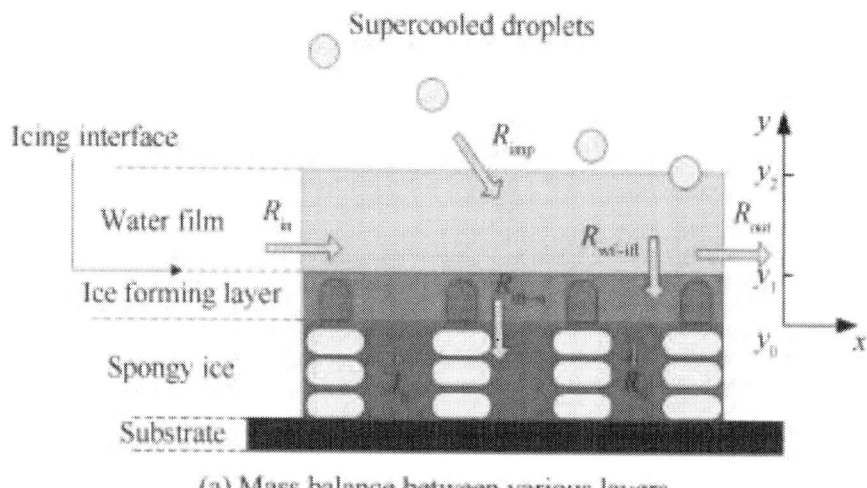

(a) Mass balance between various layers

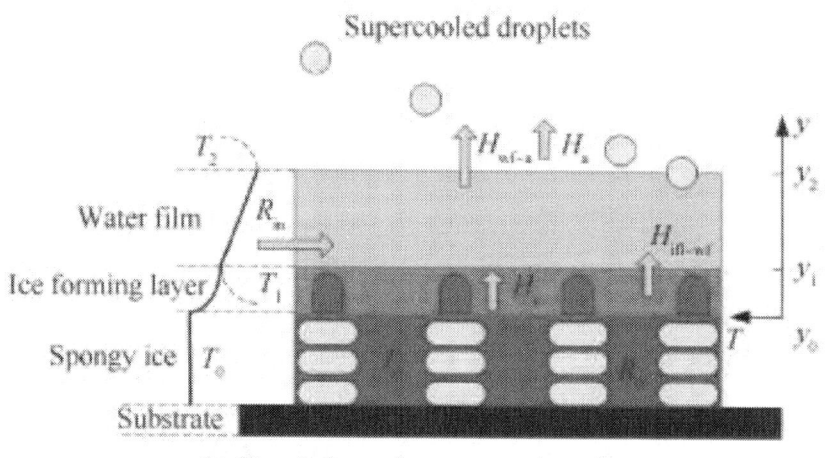

(b) Heat balance between various layers

Figure 3. Schematic of the model's surficial structure.

The conduction and evolution of heat of the ice forming zone are described by a 1-D steady-state diffusion equation, known as

$$\frac{d}{dy}\left(k\frac{dT(y)}{dy}\right) + q_v = 0 \tag{9}$$

where y is the coordinate normal to the surface with origin $y_0 = 0$ at the interface, $T(y)$ the temperature in the ice forming layer, k the thermal conductivity of the ice forming layer, which is assumed to be independent of y, with the boundary condition $T = 0$ °C and $dT/dy = 0$ when $y = 0$. The solution for Eq. (9) is

$$T_0 - T_1 = q_v(y_1 - y_0)^2/2k \tag{10}$$

The volumetric rate of evolution of latent heat in the ice forming layer can be written as

$$q_v = \frac{q_1}{y_1 - y_0} \tag{11}$$

where q_1 is the latent heat from the ice forming layer to the laminar layer, which can be expressed by

$$q_1 = I_0 L_f - c_w(T_0 - T_1)(I_0 + R_0) \tag{12}$$

where c_w is the specific heat capacity of pure water at 0 °C(4.2×10^3 J $\cdot$ kg^{-1} $\cdot$ K^{-1}), and L_f is the specific latent heat of fusion of pure water at 0 °C(3.34×10^5 J $\cdot$ kg^{-1}).

The crystal growth theory pointed out that the growth rate at the tips of dendrites is a function of the temperature difference of the bulk liquid into which the dendrite is growing[19]; the growth rate can be expressed as

$$V_c = a\Delta T^b \tag{13}$$

where a and b are empirical coefficients and ΔT is the temperature difference, $\Delta T = T_0 - T_1$. Like List, [13] we assume that the growth rate of the icing surface is a function of the temperature drop across the ice forming layer, Tirmizi recommended values a and b of 1.87×10^{-4} and 2.09. However, in order to consider the influence of the ambient temperature and to apply this model to aircraft icing field successfully, after calibration work, we recommend b as of $-3.79 \times 10^{-4}T_a^3 - 1.843 \times 10^{-2}T_a^2 - 0.15406T_a + 1.6569$, where T_a is the temperature of air flow and b a sensitive function of the temperature, which has a decisive effect on the speed of the icing interface, and also plays an important role in dividing the type of ice accretion.

Assume V_c to be the rate of advance of liquid across the icing interface, which means

$$V_c = V_1 = \frac{I_0 + R_0}{\rho_w}$$
(14)

To determine the thickness of the ice forming layer is very important for modeling the ice forming layer. Tirmizi assumed that there is a continuous linear relationship between temperature drop and the radius of curvature of the tip of a growing ice dendrite, such as

$$r_c = c + d/\Delta T$$
(15)

Tirmizi recommended applying $c = 6.16 \times 10^{-5}$ and $d = 2.024 \times 10^{-5}$ for dendritic ice, but for aircraft icing we recommended applying $c = 4.36 \times 10^{-5}$ and $d = 2.864 \times 10^{-5}$. The main function of the two parameters is to make sure that the iterative solution of the spongy icing model is convergent.

The thickness of ice forming layer is proportional to the radius of freely-growing ice dendrite tips

$$y_1 - y_0 = k_r r_c$$
(16)

Where k_r is the factor of proportionality; for aircraft icing, recommend $k_r = 1.32$. The main function of k_r is not only to guarantee iterative solution of the spongy icing model is convergent but also making sure the predicted ice shapes have physical significance. K_r is dependent of velocity and liquid water content. In recent years, a wide variety of tests have been performed in the NASA Lewis icing research tunnel (IRT). [22] These data facilitates a systematic calibration of the parameters mentioned above, and the parameters are calibrated by experimental data of different temperatures, liquid water contents, MVD, and different speeds. After the calibration work, c and d are constant, no matter for the rime, the glaze or the spongy regime, but b depends on ambient temperature, for b is a function of temperature. From Eqs.(15) and (16) we can conclude that fine ice crystals are (grow at high film supercoolings) with tips of smaller radius of curvature than coarser ice crystals (grow at lower supercoolings). Experiments have proved that high temperature (low film supercoolings) produced coarse ice columnar crystals and vice versa. [20]

Heat balance of laminar layer film

The heat balance for the laminar layer is

$$H_{in}+H_{ifl-wf}+H_{wf-a}+H_{out}=0 \tag{17}$$

Where H_{in} is the sensible heat flux in the laminar layer which is used to heat the inflowing liquid, and H_{ifl-wf} the bulk heat flux from the ice forming layer to the laminar layer, the initial value of H_{out} is zero, and the value will be updated during iteration process. H_{wf-a} is the heat flux from the laminar layer to the air flow.

The expression for the H_{in} is

$$H_{in} = C_w R_{in} \left(T_{in} - \frac{T_1 + T_2}{2} \right) \tag{18}$$

where T_{in} is the mean temperature of the mass flux into the control volume, $(T_1 + T_2)/2$ the mean temperature of the control volume under consideration, and C_w the specific heat of pure water at 0 °C.

The heat flux exported from the ice forming layer into the laminar layer is

$$H_{ifl-wf} = I_0 L_f + C_w R_{wf-ifl} \left(\frac{T_1 + T_2}{2} - T_0 \right) \tag{19}$$

The bulk heat flux between the laminar and airstream is

$$H_{wf-a} = -k_w \frac{T_1 - T_2}{y_2 - y_1} - C_w R_{wf-ifl} \left(\frac{T_1 + T_2}{2} - T_2 \right) \tag{20}$$

where k_w is the thermal conductivity of water (0.58 W · m^{-1} · K^{-1}).

The energy balance for the outer surface of the laminar film is

$$H_c + H_a = 0 \tag{21}$$

where H_a is the heat flux on the outer surface of the laminar film, and H_c the conductive heat flux directed through the laminar layer into airstream from the ice forming layer, which is a component of H_{wf-a}.

The conductive heat flux is

$$H_c = k_w \frac{T_1 - T_2}{y_2 - y_1} \tag{22}$$

H_a can be expressed as follows (more details can be found in Ref. [23]):

$$H_a = -h_{cv}(T_2 - T_a) - \frac{\varepsilon h_{cv} L_v}{c_p P_a} \left(\frac{Pr}{Sc}\right)^{0.63} (e_3 - RHe_a) - \sigma a(T_2 - T_a)$$
$$- c_w R_{imp}(T_2 - T_{imp}) \tag{23}$$

where h_{cv} and L_v (2.26×10^6 J $\cdot$ kg^{-1}) are convective heat transfer coefficient and the specific latent heat of vaporization respectively, e_3 and e_a represent the saturation vapor pressures at temperatures T_a and T_3 respectively; σ is the Stefan–Boltzman constant (5.67×10^{-8}), ε the ratio of the molecular weights of water and dry air (0.622), and r_a linearization constant for thermal radiation (8.1×10^7). RH is the relative humidity of the air, Sc is Schmidt number (0.7), and Pr is Prandtl number.

The definition of ice fraction can be written as

$$f = \frac{I_0}{I_0 + R_0} \tag{24}$$

k is the function of k_w and k_i:

$$k = k_i f + k_w (1-f) \tag{25}$$

where k_i is the thermal conductivity of ice (2.25 W $\cdot$ m^{-1} $\cdot$ K^{-1}).

ICING REGIMES

Once the thermal profile of surficial layers is known, the growth regime can be determined. Conventionally, ice shapes are generally classified as glaze, mixed and rime accretions. As air temperature rises, the spongy icing model identifies icing conditions that include the rime accretion, the spongy without water film, the spongy with water film, and the glaze accretion. The outer surface temperature of the laminar layer, T_2 is determined by the heat balance equation for the outer surface of the laminar layer. If the calculated flux of entrapped liquid satisfies the condition, $R_0 = 0$, then the glaze icing regime will occur, i.e. both f and R_0 are equal to zero. The spongy with water film regime is predicted if the impinging mass flux of water is large enough to form both a water film of excess liquid and a spongy accretion, i.e. $R_{out} > 0$. As the temperature drops, the formation rate of pure ice becomes larger (Eq. (13)). If the impinging flux, R_{imp} is large enough to form a spongy accretion with no excess liquid ($R_{out} = 0$), and the formation rate of pure ice is less than the impinging flux (i.e. $I_0 < R_{imp}$ but $I_0 + R_0 > R_{imp}$), in this case the spongy without water film regime occurs. If T_2 is below 0 °C and the formation rate of pure ice is more than the flux of impinging droplets, no liquid will be entrapped and the ice growth rate can be determined by the flux of impinging droplets, and rime regime occurs.

FURTHER STUDY OF SPONGY ICING MODEL

In Messinger icing model, one assumption is that the temperature of interface between the liquid and the ice is 0 °C, and the interface is infinitely small. Messinger model can predict rime ice, mixed ice and glaze ice, when glaze ice occurs, the surface temperature equals 0 °C. But in the present model the spongy ice layer, the icing interface and the water film must be supercooled in order to drive the conductive heat away from the icing interface towards the airstream, which has been concluded in experiments.[13] When there is no water entrapped in the spongy ice layer, the ice fraction equals 1. In rime and glaze cases, the ice fraction equals 1. For the present model, the growth rate of the ice matrix will not be less than the rate of the supercooled liquid through the icing interface:

$$V_0 = \frac{I_0}{\rho_i} + \frac{R_0}{\rho_w} \geq V_c = a(\Delta T)^b = \frac{I_0 + R_0}{\rho_w}$$

(26)

In both rime or glaze ice cases, $R_0 = 0$ and we can substitute Eq. (11) into Eq. (10):

$$T_0 - T_1 = \frac{q_1(y_1 - y_0)}{2k} = \frac{(I_0 L_f - C_w I_0 \Delta T)(y_1 - y_0)}{2k} \tag{27}$$

Substituting Eqs. (15) and (16) into Eq. (27):

$$I_0 = \frac{2k\Delta T}{k_r(c + d/\Delta T)(L_f - C_w \Delta T)} \tag{28}$$

Substituting Eq. (28) into Eq. (26), we have the following equation:

$$-C_w c \Delta T^b + (L_f c - C_w d)\Delta T^{b-1} + L_f d \Delta T^{b-2} - \frac{2k}{ak_r \rho_i} \leqslant 0 \tag{29}$$

Then, we can solve the above equation. For example, when $t_a = -4.4\,°C$, it can be solved as $\Delta T \leqslant 0.743\,°C$, which means when $t_a = -4.4\,°C$, glaze regime occurs, and the icing interface temperature will be hold in this range $0 > T_1 \geqslant -0.743\,°C$.

RESULTS

In Fig. 4 and Fig. 5, the water collection efficiency β is plotted vs surface distance s in mm. The surface distance is measured from the nose (a reference point where $s = 0$ mm) corresponding to the location where y equals zero. For the clean airfoil, the nose is at the leading edge; for the iced airfoils, the nose is located between the ice horns. Note that negative surface distance corresponds to the upper surface of the airfoil.

Analysis of small and large droplets impingement data for all geometries tested are presented in this section. The analysis impingement data are obtained with present code and LEWICE. Fig. 4, Fig. 5 and Fig. 6 show the analysis data for the clean NACA23012 airfoil and the 22.5 min iced NACA23012 airfoil at 2.5° of AOA. Also shown are the experimentally measured values reported in Ref.[16].

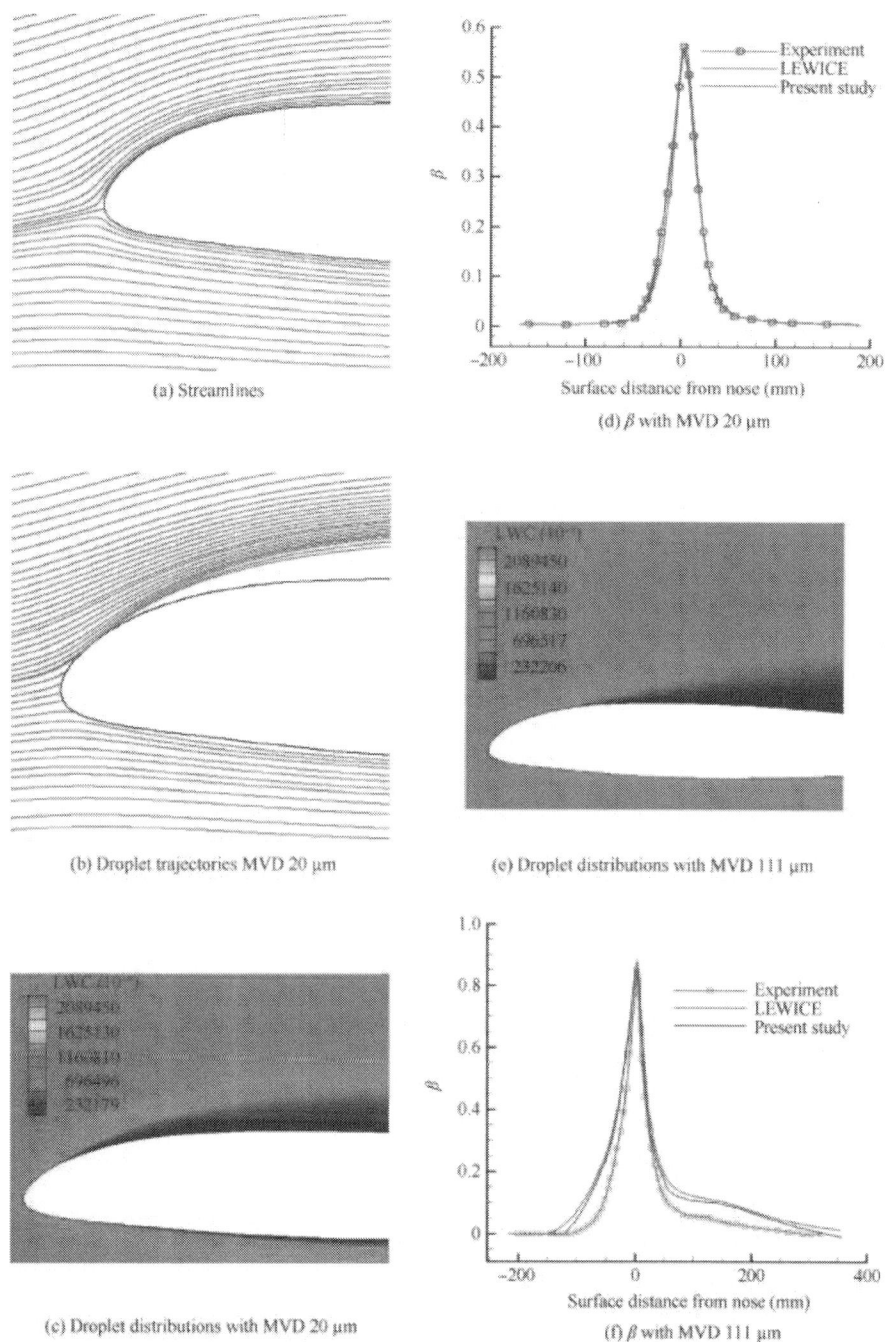

(a) Streamlines

(d) β with MVD 20 μm

(b) Droplet trajectories MVD 20 μm

(e) Droplet distributions with MVD 111 μm

(c) Droplet distributions with MVD 20 μm

(f) β with MVD 111 μm

Figure. 4. Numerical and experimental results for the clean NACA23012 airfoil.

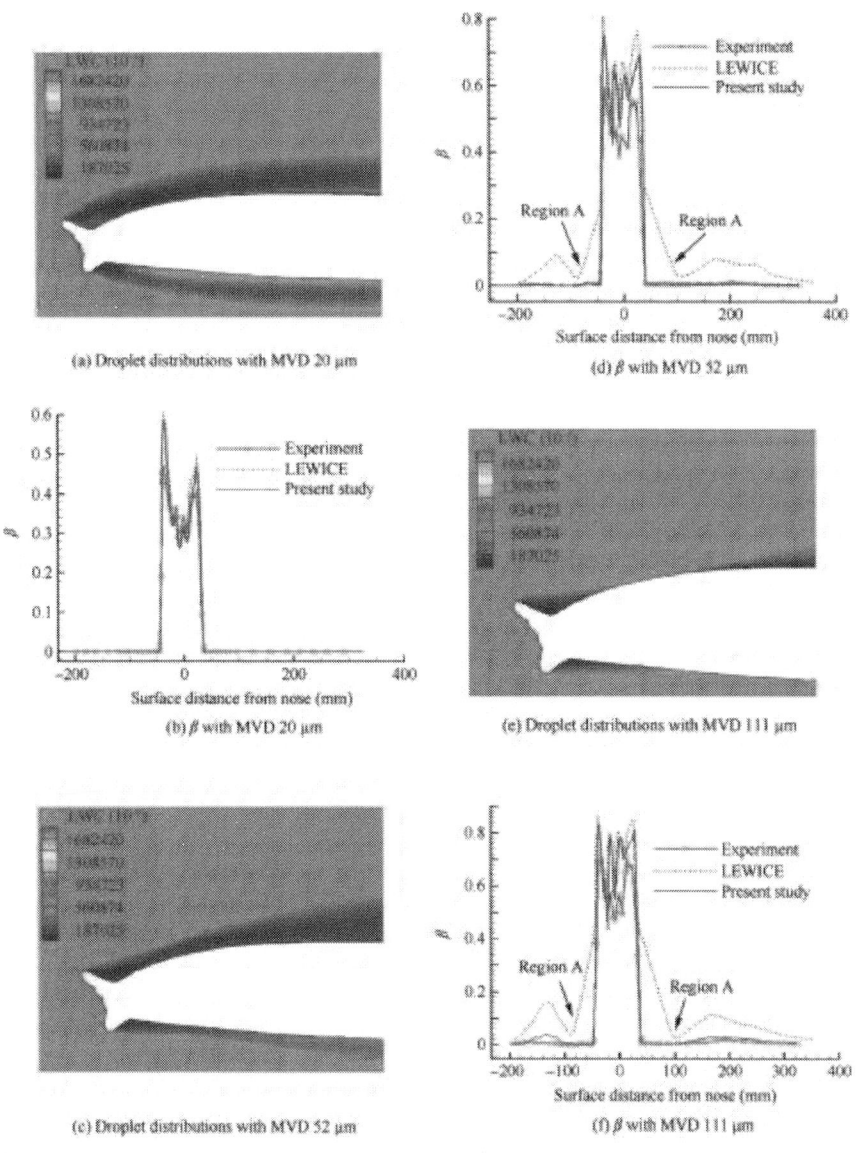

Figure 5. Numerical and experimental results for the iced NACA23012 airfoil.

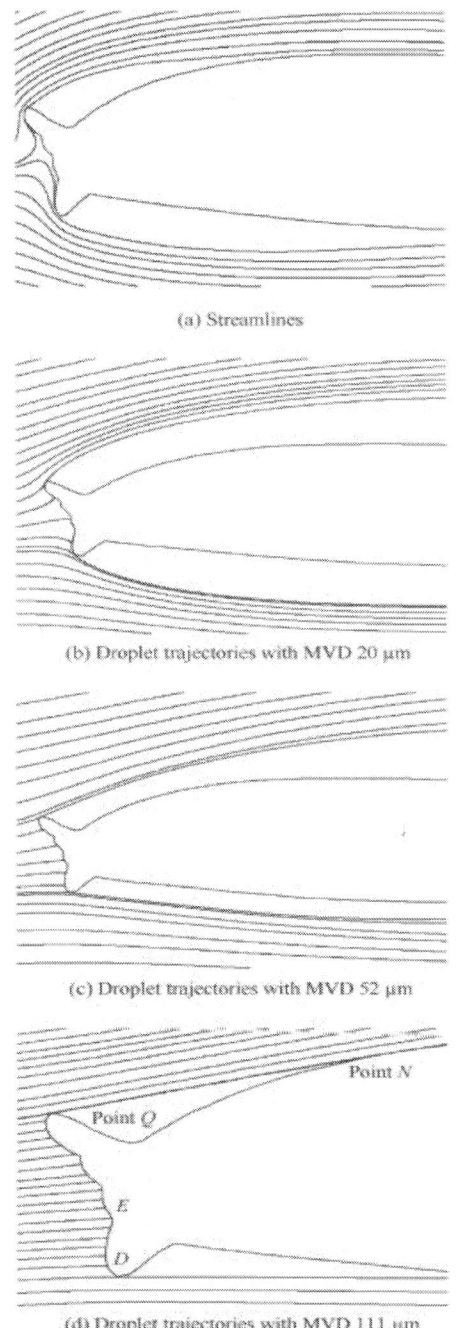

(a) Streamlines

(b) Droplet trajectories with MVD 20 μm

(c) Droplet trajectories with MVD 52 μm

Point N

Point Q

E

D

(d) Droplet trajectories with MVD 111 μm

Figure 6. Streamlines and droplet trajectories for iced NACA23012.

Impingement on clean NACA23012 airfoil

A Lagrangian particle tracking code usually computes droplets trajectories and uses them to calculate the water collection efficiency. However, for the Eulerian approach, the trajectories are not required for the computation of the collection efficiency. However, from the trajectories, one can tell the impingement limits easily. Fig. 4(a) and (b) demonstrate the streamlines and droplet trajectories for the clean NACA23012 airfoil with MVD = 20 μm. From the streamlines and the trajectories, it can be seen that droplets hit the leading edge due to the inertia, while the flow streamlines have the tendency to avoid the droplets hitting on the airfoil. For the case of MVD = 20 μm, the agreement between the experimental and CFD results is good. For the case of larger MVD = 111 μm, however, both the present code and the LEWICE code predict greater impingement limits and higher water collection efficiency compared to the experimental data. The possible reason is that the present code and the LEWCIE code do not account for the splashing effect, which is very common especially when the MVD is greater than 40 μm. It should be noted that the collection efficiency solved by the Eulerian method is lower than the collection efficiency solved by the Lagrangian method, especially around the nose of the airfoil ($s = 0$).

Impingement on NACA23012 with 22.5 min glaze ice shape

For the case of the 22.5 min ice shape and the large MVDs (52 and 111 μm), the LEWICE results in the downstream region of the horns (Region A, Figs. 5(d) and (f)) show a gradual decrease in β (collection efficiencies) compared to a sharp drop in the experimental and the present study results. The reason for this is due to the interpolation scheme used in the Lagrangian method. Generally, the interpolation scheme is fine; however, it has difficulties for geometries with multiple impingement regions which can occur on multi-element wings, highly cambered wings, and complex ice shapes. For example, one trajectory tangent with the aft impingement limit of a forward impingement region (Point Q, Fig. 6(d)) and another trajectory represents the forward limit of the aft impingement region (Point N, Fig. 6(d)); the collection efficiency between the two regions should be zero. However, for the Lagrangian method, it is not zero because of the interpolation scheme.

From Fig. 4 and Fig. 5, the results from the present study and the LEWICE code indicate higher local collection efficiency and greater impingement limits than the experimental results. There are three reasons for this. First, there is difference between the actual and the computed flow field, particularly in the region between the horns. Second, the droplet splashing is not simulated in the present code and the LEWICE code. Third, the errors are associated with the experimental investigation.

Comments on droplet trajectories

The droplet trajectories for an iced airfoil are depicted in Fig. 6 to illustrate the trends of impingement distribution in the present study. The presented droplet trajectories are for the 22.5 min ice shape with the 20, 52, and 111 μm MVD spray clouds. All trajectory computations are performed with the present code. The droplet trajectories shown in Fig. 6(b) for the 20 μm case demonstrate considerable deflection in the vicinity of the ice shape. As MVD becomes larger, the deflection of trajectories becomes progressively smaller (as shown in Fig. 6(c)); when MVD comes to 111 μm, the trajectories are practically straight (Fig. 6(d)). The reason for this is when MVD becomes larger, the inertia of the droplets becomes bigger too, so it is hard for streamlines to avoid the droplets hitting on the airfoil. Multiple local impingement peaks were observed between the ice horns. Fig. 6(d) can be used to explain how these peaks form. For example, the collection efficiency will be relatively high near Point D on the lower ice horn, for the surface is nearly normal to the incoming droplets; when the droplets hit the surface between Points D and E, the efficiency decreases due to the local slope of the surface. The reason of the peaks in the experimental results is the same as that of the peaks in the numerical results. However, the water re-impingement due to the droplet splashing may be the other reason.

Ice accretion on NACA0012

The computation is performed at four different temperatures. Fig. 7, Fig. 8, Fig. 9 and Fig. 10 show the ice shapes after six minutes of simulation from the LEWICE and the present code, which are compared to the experimental data provided by shin and bond.[4]Fig. 7shows the predicted ice shape at −19.4 °C, the ice fraction, and the mass flux of the control volume. There is no water film on the surface, and no droplet is entrapped. This is a typical rime ice. The predicted ice shape agrees well with the measured ice shape. Since it is rime and it is combined with the same Eulerian code, the spongy icing model and Messinger Model predict the same shape. Fig. 8 shows the other type of dry regime, called spongy without water film. The ice fraction holds in (0, 1), and there is no mass flux flow out of the control volume. For the temperature is relatively low, the Messinger model and spongy icing model do not have much difference, but the LEWICE predicts greater volume than the present code does. For the case of −10.0 °C (Fig. 9), this is a kind of wet ice accretion regime, named spongy with water film. The ice fraction holds in (0, 1) too, which means the liquid is entrapped in the spongy ice. The ice shape predicted by the present code with spongy icing model agrees well at suction side of the airfoil but slightly over-predicted at pressure side; while for the Messinger model, the ice shape fits not so well. There are some differences in predicted ice shapes calculated with spongy icing model and Messinger model, which is mainly due to the difference of predicted convective heat transfer coefficient. When it comes to −4.4 °C (Fig. 10), the predicted ice shape is a typical wet regime, glaze ice. The spongy icing model predicts a horn on the upper surface, which is the same as the LEWICE code does.

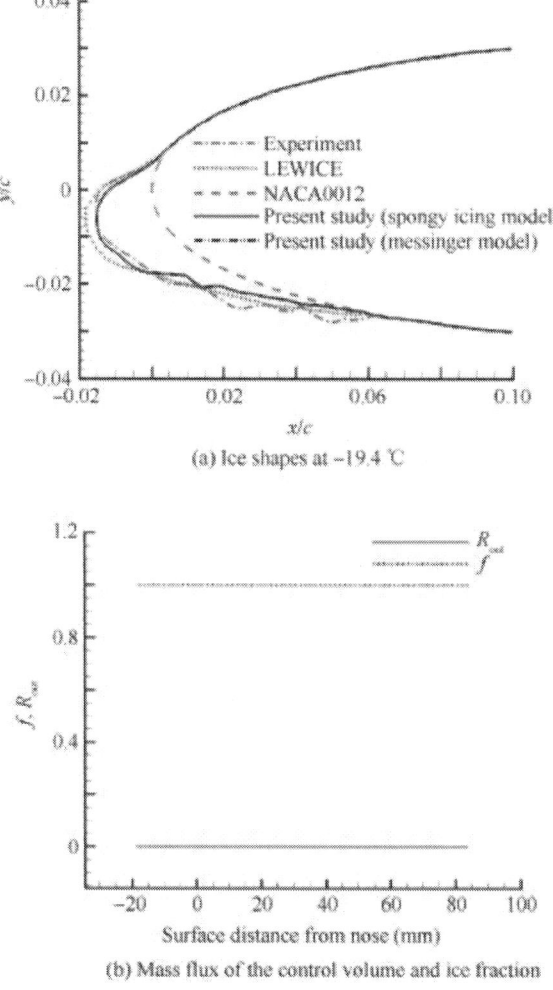

(a) Ice shapes at −19.4 °C

(b) Mass flux of the control volume and ice fraction

Figure 7. Comparison of numerical and experimental data at −19.4 °C.

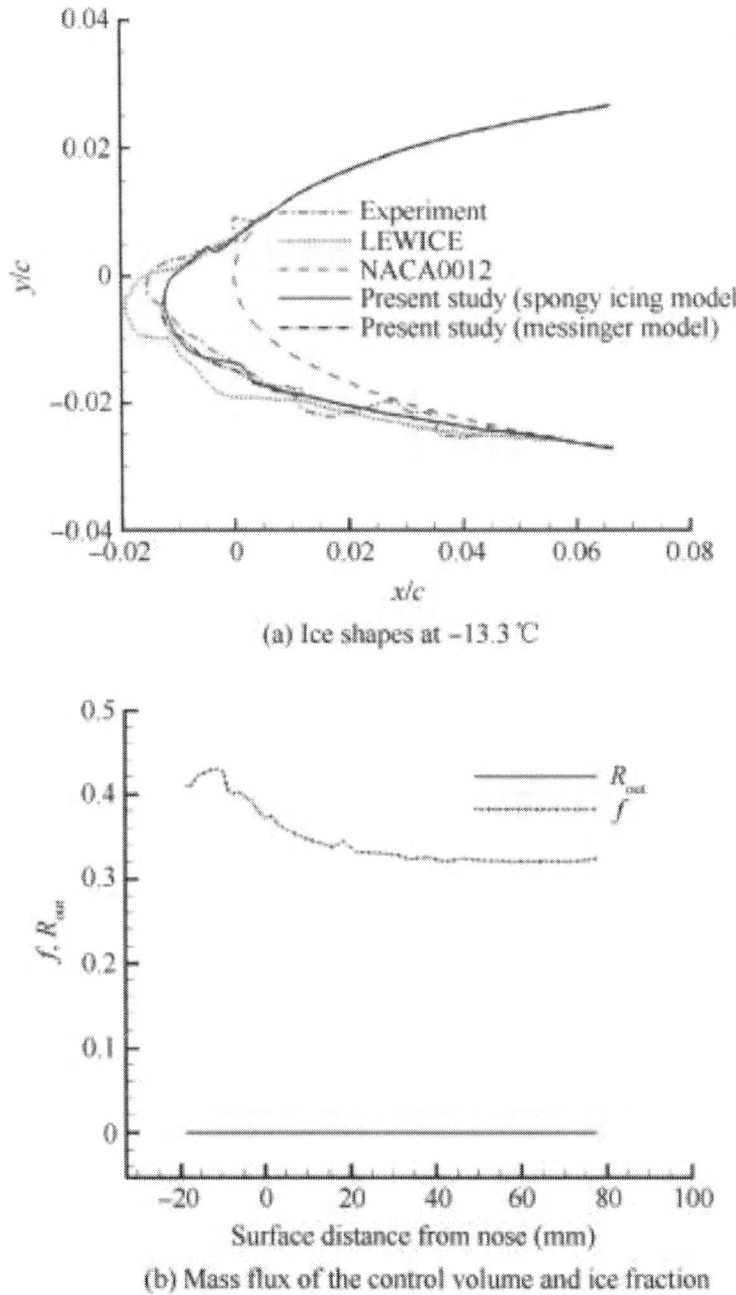

(a) Ice shapes at −13.3 ℃

(b) Mass flux of the control volume and ice fraction

Figure 8. Comparison of numerical and experimental data at −13.3 °C.

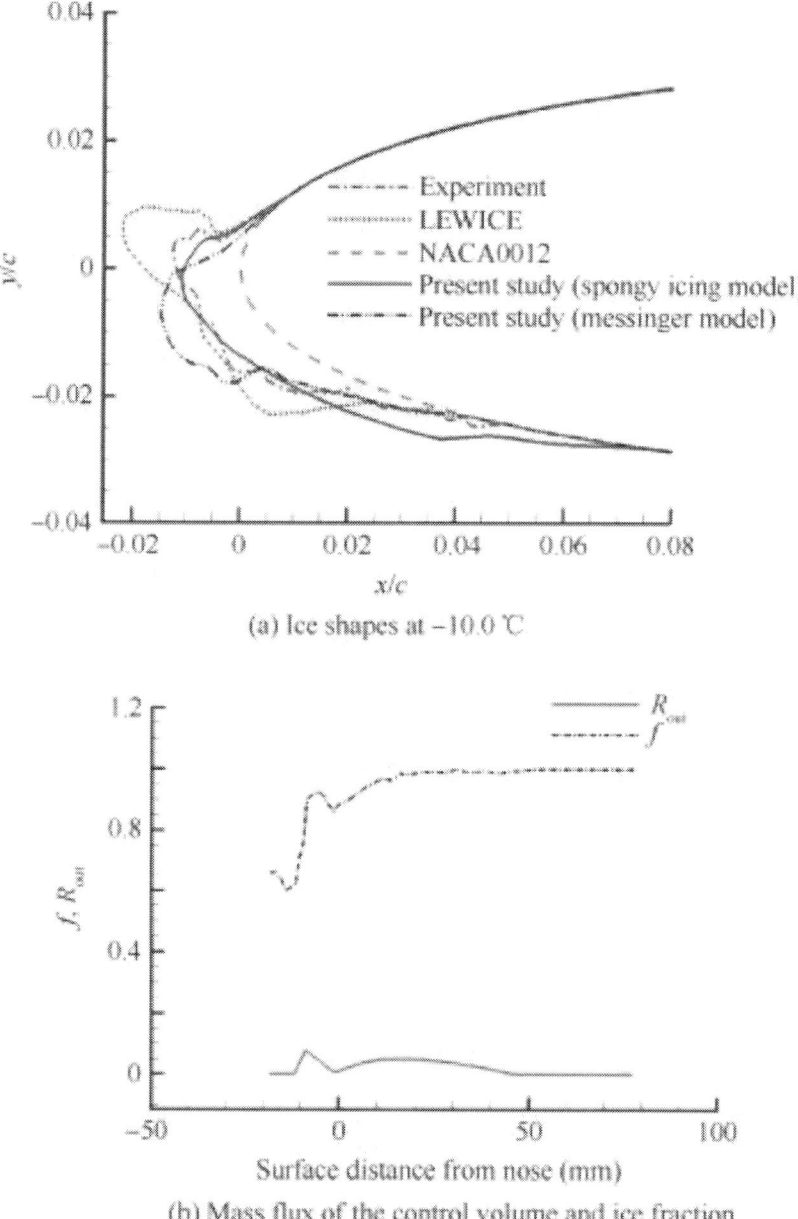

(a) Ice shapes at −10.0 °C

(b) Mass flux of the control volume and ice fraction

Figure 9. Comparison of numerical and experimental data at −10.0 °C.

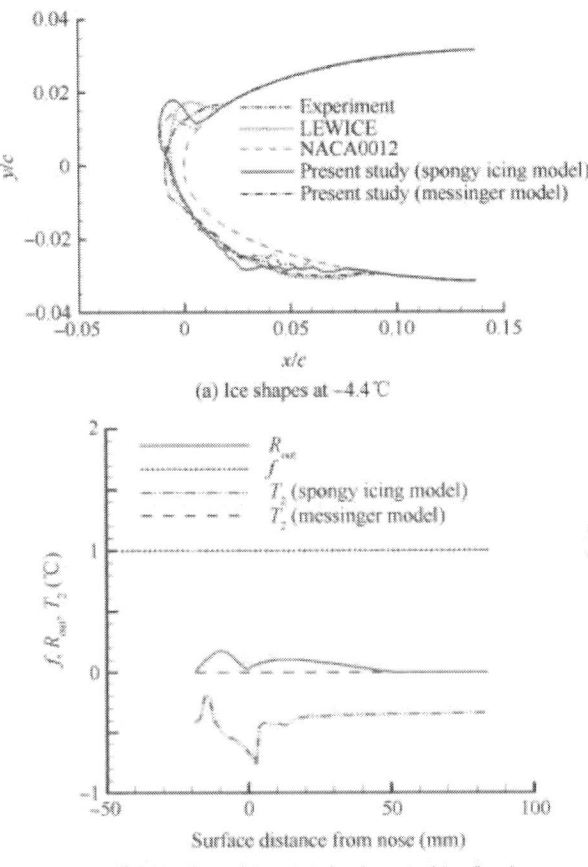

(a) Ice shapes at −4.4 ℃

(b) Mass flux of the control volume and ice fraction

Figure 10. Comparison of numerical and experimental data at −4.4 °C.

During the ice accretion process, it has been shown that the surface temperature is below the freezing point of water,[13] while it has been assumed that the temperature of the interface of ice and water film is at freezing point in Messinger model. Fig. 10(b) shows the surface temperature predicted at −4.4 °C by both models. The surface temperature predicted by Messinger model equals 0 °C while predicted by spongy icing model is below 0 °C; the temperature difference is up to 0.757 °C.

As noted above, for temperatures equal to −19.4 and −4.4 °C, corresponding to rime and glaze regime respectively. But for the other two cases (−13.3 and −10.0 °C), one is spongy without water film, another is spongy with water film.

Prediction of ice accretion and surface temperature on a cylinder

The experimental results of Ref.[14] are chosen to validate the capability of the surface temperature prediction. During his experiment, in order to measure the surface temperature of a cylinder, a non-destructive remote sensing technique was employed. Test conditions are listed in Table 3. The ice shape predicted by the present code is similar to the measured shape (Fig. 11(a)). However, the thickness near the impingement limits is under-predicted and the ice thickness at stagnation point is over-predicted.

For the surface temperature shown in Fig. 11(b), both predicted and experimental results are below 0 °C. The predicted temperature is slightly higher than the experimental data, but the trend of the surface temperature is consistent.

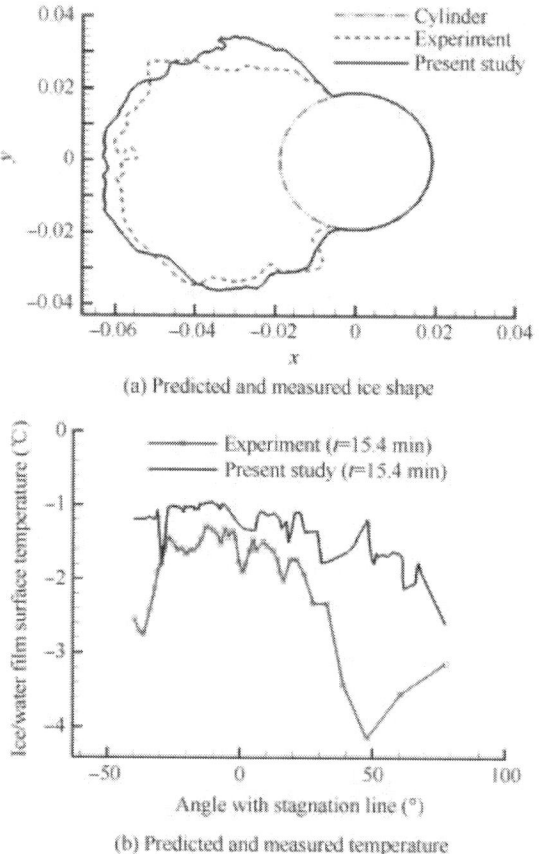

(a) Predicted and measured ice shape

(b) Predicted and measured temperature

Figure 11. Comparison of numerical and experimental data on cylinder.

CONCLUSION

A new spongy icing model based on Eulerian approach has been developed to describe glaze, spongy with water film, spongy without water film, and rime ice. A variety of experiments with different test conditions have been carried on in the NASA Lewis Icing Research Tunnel; the key parameters of the icing model are calibrated by these experimental data. The spongy icing model can be applied to aircraft icing condition, and it has a wide scope of application. The LWC range is $0-3.0$ g/m^3, the MVD range is $8.0-156.0$ µm, the velocity range is $0-200.0$ m/s, the pressure range is $47.217-101.325$ kPa and the temperature range is $-30.0-0$ °C.

To avoid numerical oscillations, a permeable wall condition is employed. Because of the numerical artifact caused by the interpolation scheme, the Eulerian approach is more feasible and accurate than the Lagrangian approach.

There are three differences between traditional models and the present icing model. The first one is the definition of the surface temperature. It has been assumed that the temperature of the interface between ice and water film is at the freezing point in traditional models, while in spongy icing model, the temperature of the interface is supercooled (e.g. when test temperature is -4.4 °C, the temperature difference is up to 0.757 °C, as shown in Fig. 10(b)). The second one is the capability of describing icing regimes. The present model can describe four different regimes while traditional models can only describe three at best. The last one is the capability of describing the sponginess of ice accretion which is observed by Fraser. The computational results are in good agreement with the experimental data. The results also point out the necessary of incorporating a droplet splash model into the present code.

ACKNOWLEDGMENTS

The authors thank professor Tom I-P Shih, head of school of Aeronautics and Astronautics, Purdue University, for giving so much help. Thanks are also due to Ph.D. Candidate Chien-Shing Lee and Research Scientist X. Chi from Purdue University for giving advices.

REFERENCES

1. Bourgault Y, Habashi WG, Dompierre J, Boutanios Z, Di Bartolomeo W. An Eulerian approach to supercooled droplets impingement calculations. AIAA-1997-0176; 1997.
2. Bragg MB. Rime ice accretion and its effects on airfoil performance [dissertation]. Columbus: The Ohio State University; 1981.
3. Shen XB, Lin GP, Bu XQ, Yu J, Hou PX. Three-dimensional analysis on ice shape of engine inlet lip. Acta Aeronaut et Asronaut Sin 2013;34(3):517–24 [Chinese].
4. Shin J, Bond T. Results of an icing test on a NACA0012 airfoil in the NASA Lewis Icing Research Tunnel. AIAA-1992-0647; 1992.
5. Bragg MB. Predicting rime ice accretion on airfoils. AIAA J 1985;23(3):381–7.
6. Potapczuk MG, Bidwell CS. Numerical simulation of ice growth on a MS-317 swept wing geometry. AIAA-1991-0263; 1991.
7. Fossati M, Khurram RA, Habashi WG. An ALE mesh movement scheme for long-term in-flight ice accretion. Int J Numer Methods Fluids 2012;68(8):958–76.
8. Messinger BL. Equilibrium temperature of an unheated icing surface as a function of air speed. J Aeronaut Sci 1953;20(1):29–42.
9. Myers TG. An extension to the messenger model for aircraft icing. AIAA J 2001;39(2):211–8.
10. Myers TG, Charpin JPF. A mathematical model for atmospheric ice accretion and water flow on a cold surface. Int J Heat Mass Transfer 2004;47(25):5483–500.
11. Bourgault Y, Beaugendre H, Habashi WG. Development of a shallow water icing model in FENSAP-ICE. J Aircr 2000;37(4):640–6.
12. Otta SP, Rothmayer AP. A simple boundary-layer water film model for aircraft icing. AIAA-2007-0902; 2007.
13. List R. Physics of supercooling of thin water skins covering gyrating hailstones. J Atmos 1990;47(15):1919–25.
14. Karev AR, Masoud F, Kollar LE. Measuring temperature of the ice surface during its formation by using infrared instrumentation. Int J Heat Mass Transfer 2007;50(3):566–79.
15. Blackmore RZ, Lozowski EP. A theoretical spongy spray icing model with surficial structure. Atmos Res 1998;49(4):267–88.
16. Papadakis M, Rachman A, Wong SC, Yeong HW, Hung KE, Bidwell CS. Water impingement experiments on a NACA23012 airfoil with simulated glaze ice shapes. AIAA-2004-0565; 2004.

17. Karadag A, Ganjoo DK, Shih TIP. A density-based method for computing incompressible multiphase flows Part 1: formulation and solution algorithm. AIAA-2000-0459; 2000.
18. Chi X, Zhu B, Shih TIP, Addy HE, Choo YK. CFD analysis of the aerodynamics of a business-jet airfoil with leading-edge ice accretion. AIAA-2004-0560; 2004.
19. Tirmizi SH, Gill WN. Effect of natural convection on growth velocity and morphology of dendritic ice crystals. J Cryst Growth 1987;85(3):488–502.
20. Lock GSH, Foster IB. Experiments on the growth of spongy ice near a stagnation point. J Glaciol 1990;36(123):143–50.
21. Dukler AE, Bergelin OP. Characteristics of flow in falling liquid films. Chem Eng Prog 1952;48(11):557–63.
22. William BW, Adam R. Validation results for LEWICE 2.0. NASA/CR-1999-208690; 1999.
23. Szilder K, Lozowski EP, Gates EM. Modeling ice accretion on non-rotating cylinders-the incorporation of time dependence and internal heat conduction. Cold Regions Sci Technol 1987;13(2):177–91.

CITATION

Xin Li, Junqiang Bai, Jun Hua, Kun Wang, Yang Zhang, A spongy icing model for aircraft icing, Chinese Journal of Aeronautics, Volume 27, Issue 1, February 2014, Pages 40-51, ISSN 1000-9361, http://dx.doi.org/10.1016/j.cja.2013.12.004.

CHAPTER 6

Drag Prediction Method of Powered-On Civil Aircraft Based on Thrust Drag Bookkeeping

Yufei Zhang[1], Haixin Chen[1], Song Fu[1], Miao Zhang[1 2], Meihong Zhang[2]

[1] School of Aerospace Engineering, Tsinghua University, Beijing 100084, China
[2] Shanghai Aircraft Design and Research Institute, Shanghai 201210, China

ABSTRACT

A drag prediction method based on thrust drag bookkeeping (TDB) is introduced for civil jet propulsion/airframe integration performance analysis. The method is derived from the control volume theory of a powered-on nacelle. Key problem of the TDB is identified to be accurate prediction of velocity coefficient of the powered-on nacelle. Accuracy of CFD solver is validated by test cases of the first AIAA Propulsion Aerodynamics Workshop. Then the TDB method is applied to thrust and drag decomposing of a realistic aircraft. A linear relation between the computations assumed free stream Mach number and the velocity coefficient result is revealed. The thrust losses caused by nozzle internal drag and pylon scrubbing are obtained by the isolated nacelle and mapped on to the in-flight whole configuration analysis. Effects of the powered-on condition are investigated by comparing through-flow configuration with powered-on configuration. The variance on aerodynamic coefficients and pressure distribution is numerically studied.

INTRODUCTION

The powered-on nacelle has a significant effect on aerodynamic characteristics of civil aircraft. The engine jet could reduce the pressure on the wing's lower surface[1] and induce interference drag. The shock location of the wing on its upper surface might be changed by the powered-on nacelle[2] and increase wave drag. The existence of the airframe could also alter the performance of the engine, introducing additional thrust loss. In the civil aircraft design practice, the thrust drag bookkeeping (also called thrust drag accounting, TDB) procedure is necessary to decompose the thrust of the exhaust system and the drag of the airframe and to point out the source of interference.[3] Usually, the airframe and engine are designed and manufactured by different companies in the modern aircraft industry. When integrated, the interference between exhaust system and airframe could induce significant drag.[4] Accurately and rationally predicting and decomposing the performances of the airframe and engine, as well as their interference effect, is important to improve the propulsion/airframe integration design, and on the other hand, split the responsibilities and contributions of the two sides.

Flight test[5] and turbofan powered simulator (TPS)[6, 7 and 8] are reliable methods for evaluating the interference drag and do TDB for realistic civil aircraft. The flight test can only be used after the aircraft is produced and the cost is very expensive. The TPS test for full aircraft configuration is also expensive. And the TPS test conditions, such as the fan pressure ratio and the Reynolds number, sometimes may not be completely consistent with the real fight condition.

The nozzle internal drag is a main source of thrust loss and is a primary issue of TDB. The quantity of the nozzle internal drag is about 1.5%–2.0% of the engine thrust.[9] It is a big value as it is equivalent to about 15–20 drag counts of drag coefficient (1 drag count = 0.0001). Interference among the engine jet, pylon and the wing also induces about 0.3%–0.9% thrust loss.

The fundamental theory of TDB is the control volume analysis method of fluid mechanics.[10] By integrating the momentum variations of the nacelle fan nozzle and core nozzle, the internal drag of the nozzles, which can cause the thrust losses, will be obtained.[11] Two dimensionless parameters, discharge coefficient and velocity coefficient, C_d and C_v, are often used to indicate the performance of nacelle nozzle.

The accuracy of velocity coefficient is important for the nacelle thrust loss calculation. Wright's error estimation[12] shows that a 0.1% uncertainty of velocity coefficient could cause a 5% uncertainty of internal drag for a typical nacelle with a velocity coefficient around 0.98. If the velocity coefficient of the nacelle is 0.99, 0.1% uncertainty could cause a 10% uncertainty on internal drag.

In industrial applications, the nozzle internal drag is usually measured by flight simulation chamber on the ground and in static air.[9] and [13] The measured C_d and C_v results are assumed to be only varying with nozzle pressure ratio and then used directly in the flight condition, where the thrust of the exhaust system is calculated by the ideal thrust of isentropic expansion subtracting the thrust losses of the nozzles. [14]

The computational fluid dynamics (CFD) method is an appropriate method for TDB and propulsion/airframe integration in the aircraft design process. In the Boeing company, the propulsion effect had been considered even in their panel method code in the 1980s.[15] In recent years, the TDB based on CFD method is going through a rapid development in the aircraft design process, such as computing the thrust loss of nacelle chevron[16] and [17] and validating the engine efficiency.[18]

The key issue of CFD-based TDB is the prediction of the nozzle performance coefficients. Earlier results of CFD showed that[14] and [19] for three-dimensional exhaust nozzle configurations, velocity coefficient had a typical accuracy in the range of 0.5% to 1.0%. This accuracy is not quite adequate for modern aircraft design. The CFD method has received its rapid development in recent years, as the computational resources go through an explosive growth. Numerical schemes and turbulence models are also greatly improved. The American Institute of Aeronautics and Astronautics (AIAA) held the first Propulsion Aerodynamic Workshop (PAW 1) in 2012.[20] The basic objective is to evaluate and improve the state-of-art of nozzle performance prediction.[21] and [22] A series of nozzle test cases with experimental data was used for CFD verification and validation in the PAW 1.

In this paper, a drag prediction process fully based on Reynolds averaged Navier–Stokes (RANS) computation is introduced. The process is first derived by applying the control volume theory on a powered-on nacelle. CFD code's accuracy is then validated by the nozzle test cases of the PAW 1. Finally the whole method is applied to the thrust and drag decomposition of a realistic configuration of a civil aircraft.[23] Variation of the nacelle velocity coefficient and the influence of the engine jet on the wing characteristics are both investigated.

CONTROL VOLUME ANALYSIS OF A POWERED-ON NACELLE

Flow regime definition

Fig. 1 shows the control volume usually used for a powered-on nacelle.[11] The planes E0, E9 and E19 represent the far field boundaries. Plane E1 is the nacelle inlet lip surface. Plane E12 is the fan face station. Planes E7 and E17 are the core nozzle entrance (or core turban exit station) and fan nozzle entrance (or fan exit station), respectively. Planes E8 and E18 are the nozzle exit planes of the core and the fan nozzle. Such a station number designation follows the conventional way in the engine industry.[3] The V_{Pre} and V_{Post} are the pre-entry control volume and post-exit control volume.

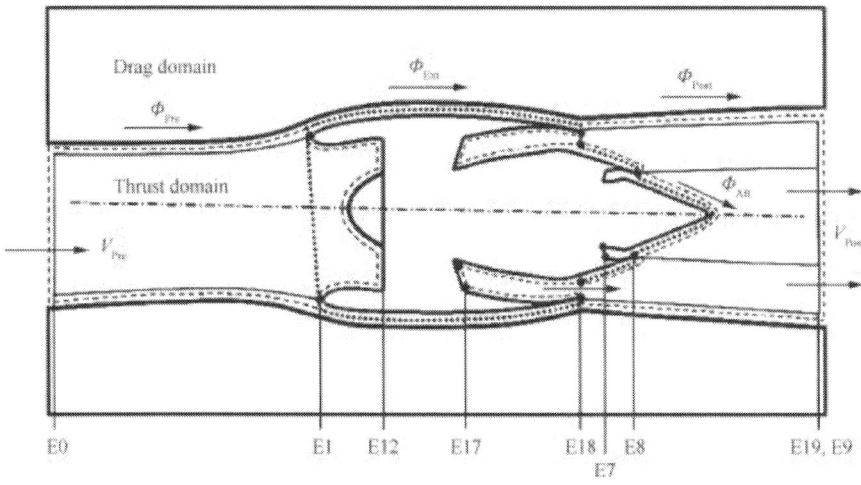

Figure 1. Sketch map of nacelle control volumes and station numbers.

The whole computation domain can be divided into drag domain and thrust domain. The inflow stream tube from plane E0 to E1 and the fan jet stream tube from E18 to E19, as well as the engine's external surface, are used as the borders separating the two domains. If the inflow stream tube is not straight, there will be a pressure force Φ_{Pre} acting on this pre-entry tube and such a force will contribute to the control volume V_{Pre}. Similarly post-exit stream tubes are coaxially formed by the core jet and fan jet, which are from E8 to E9 and E18 to E19, respectively. If the jet goes through an expansion or a contraction, the tube is not straight and a pressure force Φ_{Post} will exert effects on V_{Post}.

On the external surface of the nacelle, the summation of pressure force and viscous force is named as Φ_{Ext}. Φ_{Aft} is the resultant of the pressure and friction forces which exert on the exposed engine cowl and plug surfaces by the fan and core jets after E8 and E18 planes.

Force analysis of control volume

Define F_8 and F_{18} as the overall gross thrusts of the E8 and E18 control planes. The "thrusts" are formed by momentum forces and pressure forces. Similarly, define F_1 as the gross force on the inlet lip plane. The expressions of the forces are shown in:

$$\begin{cases} F_8 = \dot{m}_8 u_8 + (P_8 - P_{amb})A_8 \\ F_{18} = \dot{m}_{18}u_{18} + (P_{18} - P_{amb})A_{18} \\ F_1 = \dot{m}_1 u_1 + (P_1 - P_{amb})A_1 \end{cases} \tag{1}$$

where the subscripts represent the respective stations on the control volume in Fig. 1, the subscript "amb" represents the ambient air condition, $\dot{m}$ is the mass flow rate, u the flow velocity, P the pressure and A the area.

On the far field planes E0 and E9 + E19, the static pressures are equal to the ambient pressure. By the momentum balance and mass flow conservation on the control volumes V_{Pre} and V_{Post}, we can get

$$F_8 + F_{18} + \Phi_{Post} + \Phi_{Aft} = \dot{m}_8 u_9 + \dot{m}_{18}u_{19} \tag{2}$$

$$F_1 = \dot{m}_1 u_0 + \Phi_{Pre} \tag{3}$$

Here to get the Φ's, the pressure force in the integrand should be P-P_{amb} instead of P.

The net propulsive force generated by the nacelle hence should be

$$\begin{aligned} F_{NPF} &= (F_8 + F_{18} - F_1) - (\Phi_{Ext} + \Phi_{Aft}) \\ &= (\dot{m}_8 u_9 + \dot{m}_{18}u_{19} - \dot{m}_1 u_0) - (\Phi_{Pre} + \Phi_{Post} + \Phi_{Ext} + \Phi_{Aft}) \\ &= F_N - D \end{aligned} \tag{4}$$

where $F_N = \dot{m}_8 u_9 + \dot{m}_{18} u_{19} - \dot{m}_1 u_0$ and D=$\Phi_{Pre}+\Phi_{Ext}+\Phi_{Post}+\Phi_{Aft}$, F_N is called "overall net thrust", and D is the nacelle aerodynamic drag.

The above control volume locations are before E1 or after E8/E18. However in propulsion/airframe integration analysis, the engine supplier usually provides the engine's working conditions by giving the total pressure and total temperature at the E7 and E17 planes. For the inlet, the static pressure boundary condition is given at the E12 plane. From these planes to E8/E18 and E1, additional losses will be produced by the nozzle and inlet. They will be discussed in the following section.

Classification of drag sources

If the inflow stream tube expands when approaching the nacelle lip, the pressure forceΦ_{Pre} acts on this pre-entry tube will contribute to the control volume as drag. It can also be understood that the engine inlet cannot capture all the flow through the equivalent area in the far field. Hence such a drag is also called spillage drag. When their signs are correctly defined, Φ_{Pre},Φ_{Ext} can all be called nacelle external drag. They are accounted onto the aerodynamic drag.

In the nozzles, the friction, separation and shock wave happened between E7 and E8 planes or between E17 and E18 planes are producing thrust loss. Such a loss is called internal drag of nacelle, including fan nozzle internal drag (from E17 to plane E18) and core nozzle internal drag (from E7 to plane E8). Φ_{Aft} is the resultant of the pressure and friction forces which exert on the engine cowl and plug surfaces exposed in the fan and core jets after the nozzle exits (plane E8 and E18). Although these surfaces face backward, when the jets are over-expanded, the P-P_{amb} integration on them will show drags to the whole engine. Since these surfaces can be understood as the extension of the nozzle internal surfaces, the contribution of Φ_{Aft} is also counted into the nozzle internal drag.

On the post-exit stream tube, if the jet goes through a contraction or expansion outside the nozzle and the jet is not well expanded to the ambient pressure, a pressure forceΦ_{Post} might contribute as drag. Φ_{Post} reveals the loss during the jets' expansion process. Apparently, if the jet is perfectly expanded to the ambient pressure through the nozzle, Φ_{Post} should be zero. When the airframe affects the expansion, the interference will be reflected in Φ_{Post}. It is counted into the thrust loss.

However, things are different for inlet. As mentioned earlier, Φ_{Pre} will be counted as aerodynamic drag conventionally. Moreover, the drag of inlet (pressure and friction forces on the internal surfaces from E1 to E12) is

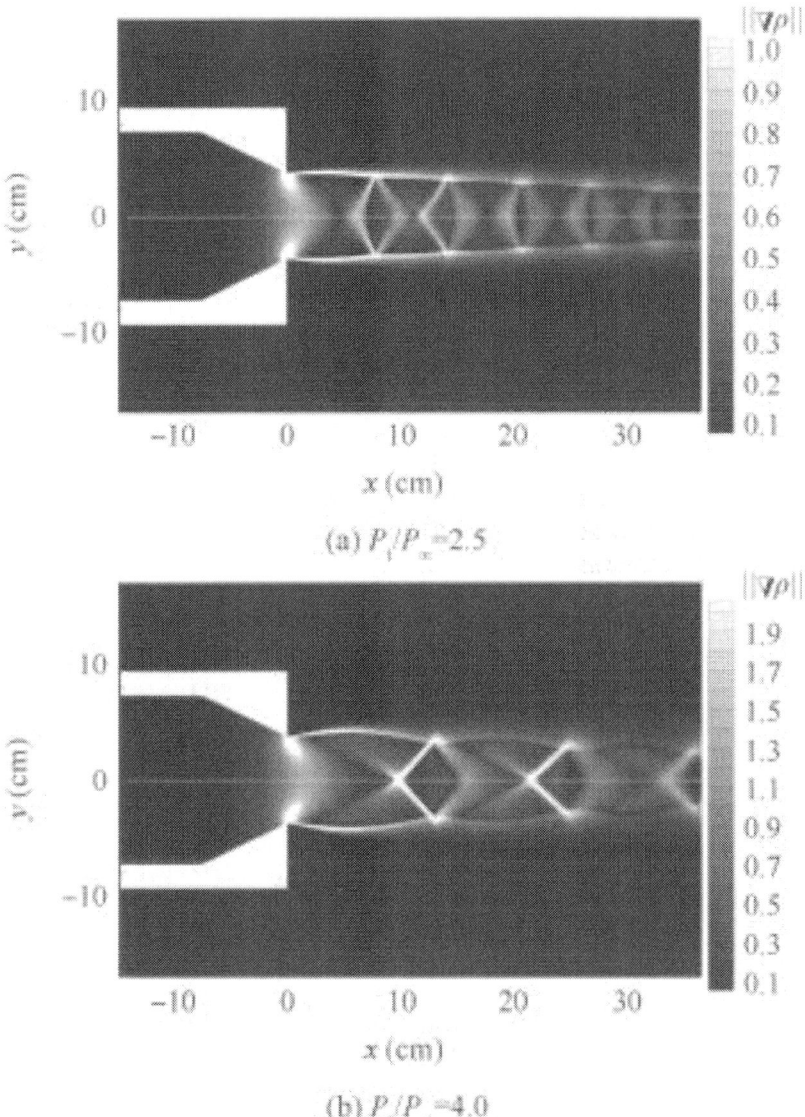

Figure 3. Typical flow field (density gradient magnitude) of 25° nozzle.

Fig. 4 shows the comparisons of the discharge coefficient and velocity coefficient. The discharge coefficient is defined as the ratio of the actual mass flow rate $\dot{m}_{actual}$ and the ideal mass flow rate $\dot{m}_{ideal}$ from the isentropic expansion, as Eq. (13), the ρ^*, v^*, A^* are the critical density, velocity and area.

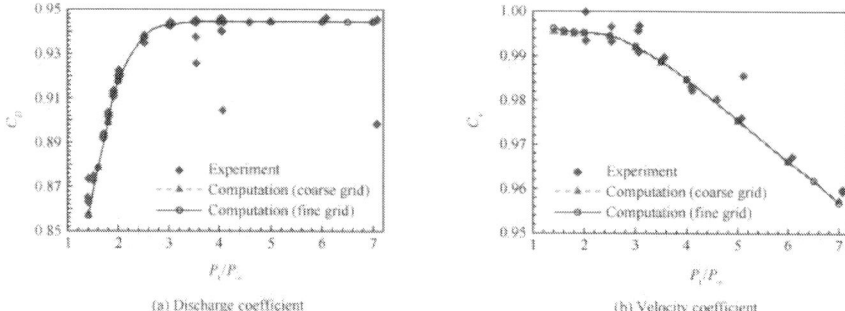

(a) Discharge coefficient   (b) Velocity coefficient

Figure 4. Discharge and velocity coefficients in the case of 25° nozzle.

The velocity coefficient of a single stream nozzle is defined as the ratio of the net thrust (when the external fluid is static) and the ideal thrust, as in Eq. (14). The ideal thrust F_{ideal} is the product of actual mass flow rate and ideal velocity. The ideal velocity can be calculated by Eq. (7). Eq. (14) is essentially identical to Eq. (8).

$$\begin{cases} C_d = \dfrac{\dot{m}_{actual}}{\dot{m}_{ideal}} = \dfrac{2\pi \int_0^{r_{exit}} \rho U r\, dr}{\dot{m}_{ideal}} \\ \dot{m}_{ideal} = \rho^* v^* A^* \end{cases}$$

$$\qquad\qquad\qquad\qquad\qquad\qquad\qquad\qquad (13)$$

$$C_v = \left.\frac{F_{net}}{F_{ideal}}\right|_{static} = \left.\frac{F_{net}}{\dot{m}_{actual} u_{ideal}}\right|_{static} = \left.\frac{Thrust - Drag}{\dot{m}_{actual} u_{ideal}}\right|_{static} \quad (14)$$

In Fig. 4, the results match well with the experimental data. Two grids are used to test the grid sensitivity. For axisymmetric computation, the coarse grid has 70 thousand points, and the fine grid has 299 thousand points. The coefficients from the two grids are nearly identical. The experimental data have uncertainty in most cases. The computation results are located within the error band of the experimental data. It seems that the CFD computation is satisfactory. A very low-speed (Mach number is 0.01) is set as the external flow speed. The effect of the external flow is discussed in Ref.[21] As the external speed is very low and the error of the external flow drag is negligible.

Fig. 5 shows the Mach number distributions of different pressure ratios at different stream wise locations, in the figure, "CFD" means computation,

"EXP" means experiment. The wall Mach number is calculated from the static pressure using the isentropic relation. In the figure, it can be seen that the flow in the nozzle keeps unchanged with the pressure ratio when the nozzle is already choked. The computation results match quite well with the experimental data. The CFD method is well validated by the test case.

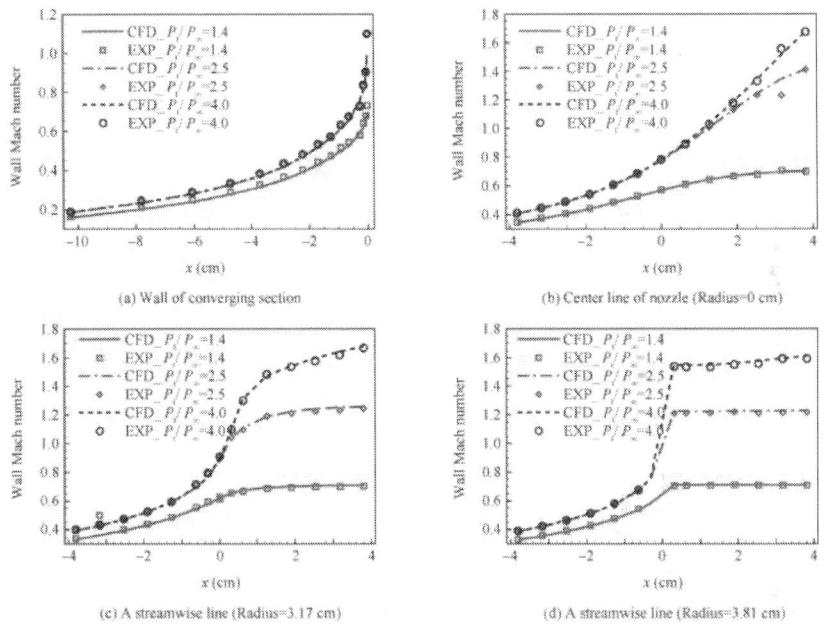

Figure 5. Mach number distributions at different locations (15° nozzle case).

DRAG PREDICTION OF A REALISTIC POWERED-ON AIRCRAFT CONFIGURATION

In this section, the drag prediction method is applied on a twin-jet civil aircraft configuration. Firstly, isolated nacelles without and with pylon are numerically investigated to determine their velocity coefficients. Then, the TDB method is used to decompose the drag and thrust of a powered-on complete civil jet configuration. Finally, the effect of the powered-on condition is analyzed by comparing the configurations with through-flow nacelle (TFN) and powered-on nacelle (PN).

C_v Computation of isolated nacelle

In order to estimate the thrust loss caused by jet scrubbing on the pylon, an axisymmetric nozzle and an isolated nacelle without and with pylon are numerically computed and compared. Fig. 6 shows the computation grids of the three configurations. The axisymmetric nozzle case, which has test data, is used to verify the computation's accuracy. The grid number is about 70 thousand on a meridian plane. The other two test cases are then computed and compared. The grid numbers of nacelle without and with pylon are 14 million and 17 million, respectively. The first layers of all the meshes in the normal direction are less than 0.005 mm to ensure that the y^- is less than 1.0.

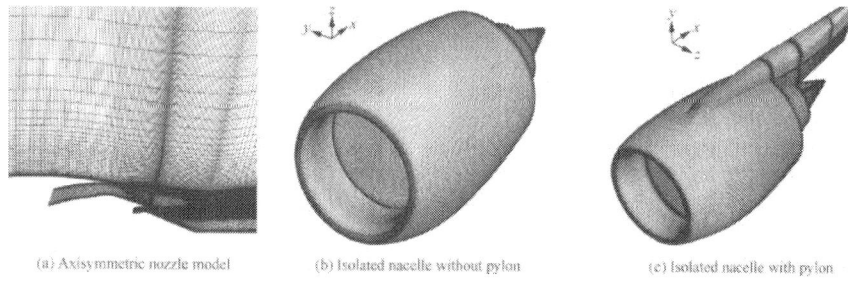

(a) Axisymmetric nozzle model   (b) Isolated nacelle without pylon   (c) Isolated nacelle with pylon

Figure 6. Computational grids of three nacelle configurations.

Powered-on condition is introduced into the nacelle computations via boundary conditions. The fan nozzle entrance and core nozzle entrance are set as total pressure and total temperature boundaries. The conditions are shown in Eq. (15). Ambient condition is air at an altitude of 11.5 km. In order to get rid of the inlet drag's influence, the fan face boundary is set as wall. Such a setting simulates the nacelle in the calibration chamber, as described in Section 2.4.

$$\frac{P_{0,\mathrm{FanNozzle}}}{P_{\mathrm{amb}}} = 2.277, \quad \frac{T_{0,\mathrm{FanNozzle}}}{T_{\mathrm{amb}}} = 1.269$$

$$\frac{P_{0,\mathrm{CoreNozzle}}}{P_{\mathrm{amb}}} = 1.604, \quad \frac{T_{0,\mathrm{CoreNozzle}}}{T_{\mathrm{amb}}} = 2.893 \tag{15}$$

In the nozzle calibration test, the external flow is static. For the compressible NSAWET code, a free stream Mach number has to be assigned to obtain a convergent solution. In the present computation, a preconditioning method is applied to making the code able to study the influence of free stream Mach number. Several Mach numbers (from 0.002 to 0.100) are computed to get the trend of the velocity coefficient.

The axisymmetric model is an idealized nozzle model for the engine design process. Preliminary test data is provided by the engine manufacturer. The CFD computation process is the same as in Section 3. Computed velocity coefficient is 99.67%, which is obtained by external Mach number 0.005, and the test result is 99.79%. Numerical error on the velocity coefficient is about 0.12%. It demonstrates that the accuracy of the present computation is satisfactory.

Fig. 7 shows the computed flow fields of the isolated nacelles without and with pylon when the free stream Mach number is 0.005. The fan nozzle jet is supersonic at the exit. Shock train can be seen in the jet wake. Although there is a very low external flow speed, the streamlines can indicate the entrained flow direction. A dead water area exists in the inlet as the fan face is set as wall boundary. Although the front half part of the nacelle (including inlet and cowl) has nothing to do with the nozzle performance and C_v, the accuracy of the drag computation will be uncertain when integrating along an unclosed surface. Hence in the present computation, the nacelle geometry is completely computed to get the nacelle C_v.

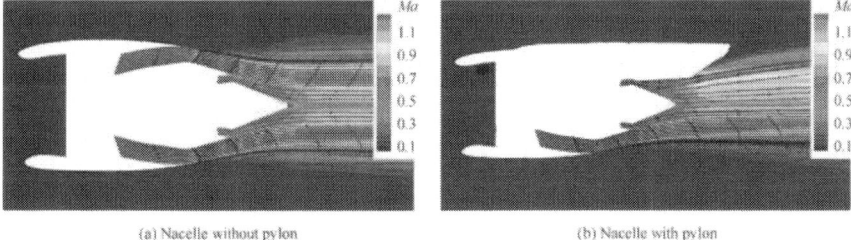

(a) Nacelle without pylon   (b) Nacelle with pylon

Figure 7. Flow field of symmetry plane of isolated nacelles ($Ma_\infty = 0.005$).

It is found that the external free stream Mach number will not change the actual mass flow. For the nacelle without pylon, the fan nozzle mass flow rate is 189.34 kg/s and the core nozzle mass flow rate is 18.65 kg/s. The ideal thrusts F_{ideal} is 71851 N. Similarly, the ideal thrust of the nacelle with pylon is 66569 N. The nozzle area reduction caused by the pylon makes the difference.

Table 1 shows the computation results of the isolated nacelles with different free stream Mach numbers. Strictly, C_v should be calculated by Eq. (8) with the free stream Mach number being zero. The results in Table

1 show that the computed F_{NPF} decreases with the Mach number increasing, because the drag on the nacelle external cowl, Φ_{Ext}, becomes significant. With a series of CFD results at different Mach numbers, Fig. 8shows the computed net thrust ratio (F_{NPF}/F_{ideal}) is almost linear with the Mach number decreasing. Since the trend of velocity coefficient with external flow approaching zero speed is clearly linear, we could extrapolate the $Ma = 0$ value of C_v with several values of bigger Mach numbers, where the computations are much easier to converge.

Table 1. Results of isolated nacelles with different free stream Mach numbers.

Free stream Mach number Ma_∞	Net thrust ratio (%)	
	Without pylon	With pylon
0 (extrapolated)	98.755	98.31
0.002	98.751	
0.005	98.745	98.301
0.01	98.736	98.291
0.02	98.717	98.272
0.03	98.699	
0.05	98.664	98.21
0.1	98.581	98.102

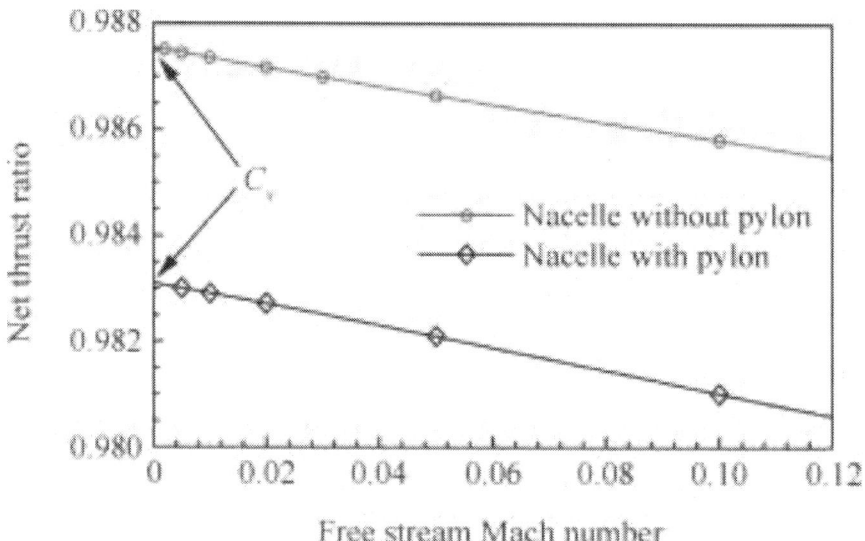

Figure 8. Net thrust ratio of isolated nacelle with different Mach numbers.

Velocity coefficients C_v are then obtained as 98.755% and 98.310% by extrapolating the lines to $Ma = 0$ for both nacelles. The interference between nacelle jet and pylon is an important drag source of powered-on configuration. The difference between the nacelle's C_v with and without pylon is introduced by the jet's scrubbing of the pylon, as well as the changing of the expansion. The former is similar to Φ_{Aft} and the later could be revealed in Φ_{Post}. Fig. 7(b) shows the flow field on the symmetry plane of the nacelle with pylon. The direction of the jet is changed by the pylon due to Coanda Effect.[30 and 31] By comparing the velocity coefficients of nacelle without and with pylon, we can see that the thrust loss caused by the pylon is 0.44%.

Application to full aircraft configuration

The drag prediction method is applied to the thrust and drag splitting in the design of a twin-jet civil aircraft. Fig. 9 shows the computation grid of the full configuration. The grid number is about 30 million for half model. The mean aerodynamic chord (MAC) is 4.25 m. The dimensionless thickness of the first grid layer in the wall normal direction is 2.0×10^{-6} normalized by the MAC. In order to study the jet's effect on the wing, both the full configurations with through-flow nacelle and powered-on nacelle are numerically investigated. The grids of the two configurations have the same grid density and distribution. Grid difference only exists near the nacelle core region.

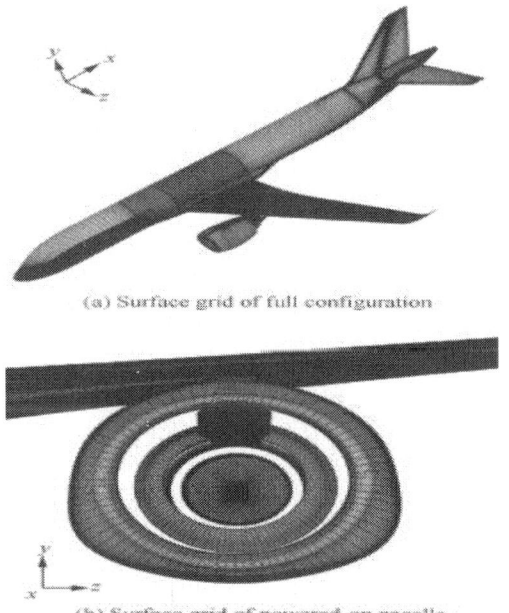

(a) Surface grid of full configuration

(b) Surface grid of powered-on nacelle

Figure 9. Computation grid of full and powered-on nacelle configurations.

The TPS experiment is complicated and expensive in industry and rarely conducted, while the experiment of full aircraft with TFN is cost acceptable and therefore routine. In the present study, the full configuration with TFN was tested in the German-Dutch Wind Tunnels (DNW-HST). The experimental result is used to validate the accuracy of the computation.

Fig. 10 shows the comparison of the wind tunnel data and computational result. The free stream Mach number is 0.2, and the Reynolds number based on the MAC is 2.8 million. The computational lift and drag coefficient curves match well with the experimental data. It shows that the accuracy of the CFD code and the grid is satisfactory. Additional cruise condition experiment of the full configuration with TFN was tested in the European Transonic Windtunnel (ETW). Under the cruise conditions of $Ma = 0.785$, $Re = 2.4 \times 10^7$, $C_L = 0.55$, the drag coefficient C_D of the experiment is 0.03090, while the computation is 0.03151. The computation result matches very well with the experiment.

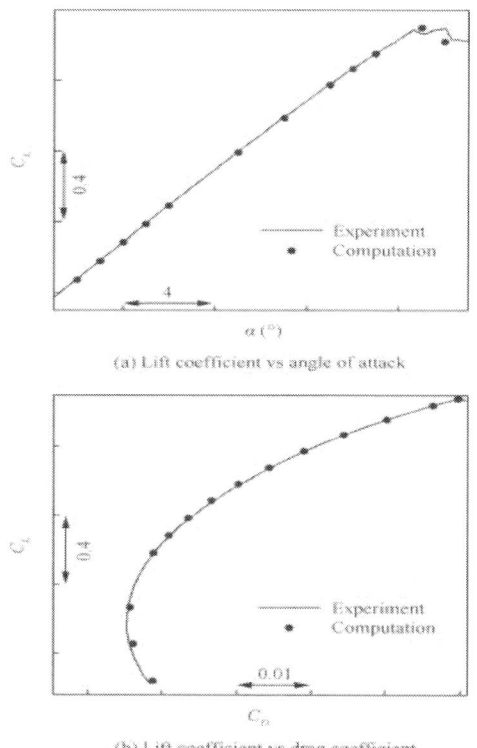

(a) Lift coefficient vs angle of attack

(b) Lift coefficient vs drag coefficient

Figure 10. Comparisons of lift and drag coefficient curves (TFN configuration, $Ma = 0.2$, $Re = 2.8 \times 10^6$).

Fig. 11 shows the Mach number contour slices of the aircraft with the PN at cruise. The powered-on boundary conditions for the jet are the same as Eq. (13). The boundary condition for the fan face is set as static pressure condition, with the pressure to be adjustable to obtain a mass flow balanced solution. The mass flow balance is an important issue for TDB. Additional spillage drag and jet-effects drag[3] will be induced if the mass flow is unbalanced. For each computation, the mass flow rate of the fan face is adjusted to be equal to the total mass flow of the fan nozzle and core nozzle. For the whole lift-drag polar curve computation, the jet power condition is fixed. However, the thrust and drag values on the polar are not balanced when varying the angle of attack.

Figure 11. Mach number slices of PN configuration at cruise ($Ma = 0.785$, $Re = 2.4 \times 10^7$, $C_L = 0.55$).

Fig. 12 shows the aerodynamic coefficient curves of the full aircraft configuration, compared between TFN and PN configurations. In Fig. 12(a), the lift coefficient of the powered-on condition is a little bit lower than the TFN configuration at the same angle of attack. The jet's effect on the wing decreases the lift coefficient. At cruise, where the C_L is equal to 0.55, the angle of attack for PN is 2.86° while 2.76° for TFN.

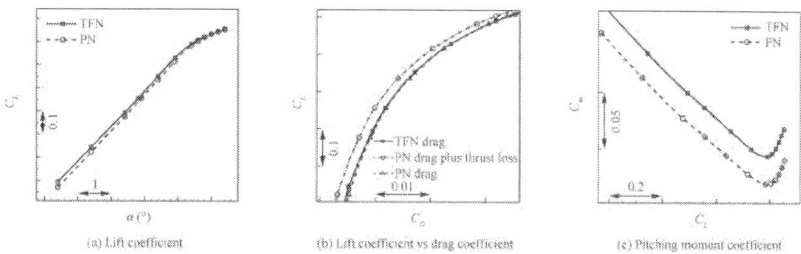

Figure 12. Lift, drag and pitching moment coefficients of full aircraft with TFN and PN.

Three curves are shown in Fig. 12(b). With the TDB procedures, two drag curves of the PN configuration are plotted. The net thrust of the engine is calculated by Eq. (10). The aerodynamic drag, which should be responsible by the airframe manufacturer, is computed by Eq. (11), in which, F_{NPF} is obtained by integrating all the pressure and friction forces on all the configuration's wall boundaries and overall gross thrust on the fan entrance and nozzle exits.

In Fig. 12(b), PN configuration's aerodynamic drag plus the thrust loss, computed by Eq.(9), is quite close to and a little higher than the TFN configuration's drag, because the nacelle's internal drag is already included in the TFN configuration. At the cruise C_L, the numbers for PN and TFN are 0.03167 and 0.03151, respectively. If the TFN configuration is accurately designed to reproduce the mass flow and thrust loss of the powered-on condition, such a difference can be treated as the interference drag caused by the jet effects. Anyway, such a difference can be used as the correction when the TFN data is used to predict the aircraft's powered-on performances.

Compared to the drag, the jet effects on the pitching moment is much more obvious. The nose-down pitching moment coefficient C_m of PN is increased a lot as shown in Fig. 12(c).

Table 2 shows the results of the PN configurations at cruise. The PN's aerodynamic drag is 0.02978. The thrust losses produced by the nozzle and pylon scrubbing are computed by the velocity coefficients pre-computed in Section 4.1 with the Eq. (9). From the difference between the C_v's with and without pylon, the pylon caused drag can be computed as 4.9 counts at cruise. It could be reduced by redesigning the pylon contraction section of the jet scrubbing region.

Table 2. Results of PN configuration at cruise ($Ma = 0.785$, $Re = 2.4 \times 10^7$, $C_L = 0.55$).

Angle of Attack	C_L	C_D (airframe aerodynamic drag)	C_D (plus thrust loss)	Total thrust loss	Thrust loss of nozzle	Pylon caused thrust loss, or drag
2.86	0.55	0.02978	0.03167	0.00189	0.0014	0.00049

The computation of thrust loss highly depends on the velocity coefficient. Fig. 13 shows the relation between the thrust loss and the velocity coefficient. If the uncertainty of the velocity coefficient is 0.1%, the error of the thrust loss is about 1.1 counts. It supports the

Wright's[12] error estimation result, for which the velocity coefficient of nacelle is about 98.3%. In the present study, the computational error of velocity coefficient compared with test data is less than 0.15%, as shown in Section 4.1. Consequently, the uncertainty of thrust loss is less than 2 drag counts. It is a reasonable value for aircraft design, as the fourth and fifth Drag Prediction Workshops show that the drag errors between either CFD and experiment, or different experiments are about 5 to 10 drag counts.[32] and [33]

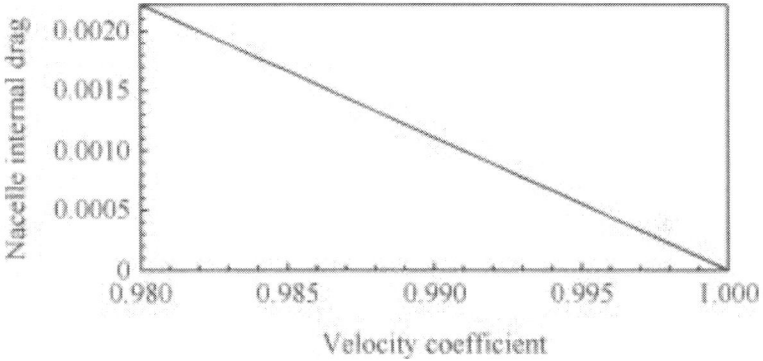

Figure 13. Relation of nacelle internal drag and velocity coefficient.

Fig. 14 shows the surface pressure coefficient C_p distribution and the streamlines of the nacelle and pylon region. A suction peak can be seen on the pylon of the PN configuration. The streamline on the pylon is changed by the jet scrubbing.

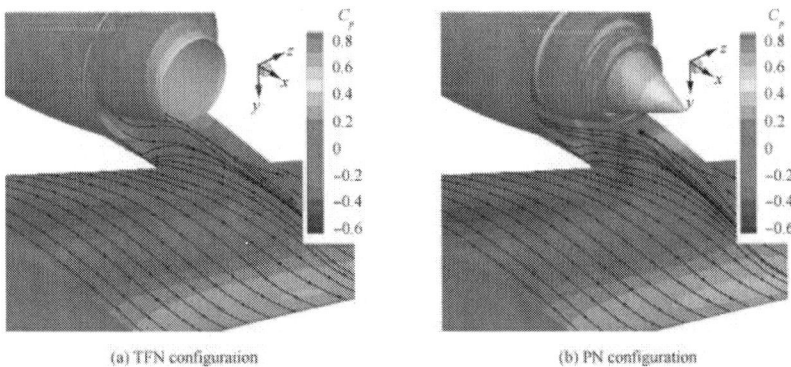

(a) TFN configuration   (b) PN configuration

Figure 14. Surface pressure and streamline of nacelle and pylon region ($Ma = 0.785$, $Re = 2.4 \times 10^7$, $C_L = 0.55$).

Fig. 15 shows the pressure distributions of the TFN configuration and the PN configuration. In this figure, z is the spanwise location, b is the span length, and c is the local chord length at the spanwise location. The engine jet has significant effects on the lower surface of the inboard wing. The negative pressure of the lower surface is increased from 20.0% to 33.7% semi-span. A suction peak appears at the lower surface of the 33.7% semi-span location, which is adjacent to the pylon. Similar phenomenon can be seen near the pylon of the wing/nacelle/pylon configuration with a long cowl nacelle.[34] The aft loading of the wing is increased by the jet, as shown in the 33.7% to 46.0% sections. This effect could increase the pitching moment. The shock locations are slightly changed by the jet effect.

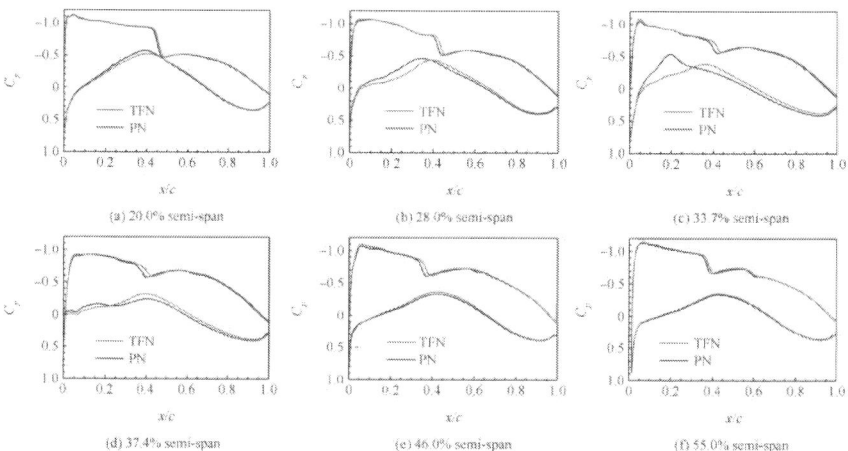

Figure 15. Pressure distribution of TFN configuration and PN configuration ($Ma = 0.785$, $Re = 2.4 \times 10^7$, $C_L = 0.55$).

CONCLUSIONS

A drag prediction method for powered-on aircraft based on thrust drag bookkeeping is introduced in this paper.

(1) A drag prediction method based on TDB is derived from the control volume theory. The process of TDB is introduced fully based on CFD. The critical problem of TDB is to accurately predict the velocity coefficient of the nacelle.

(2) The CFD's accuracy on predicting the performance coefficients is well validated by the test cases of the AIAA PAW 1. The velocity and discharge coefficients match well with the experimental data.

(3) For the TDB of an aircraft, the velocity coefficient can be computed by setting the isolated nacelle in stationary ambient air. For compressible CFD solver, C_v can be achieved by linear extrapolation from values with ambient air flowing at small Mach numbers.

(4) The aerodynamic characteristics of a powered-on twin-jet civil aircraft configuration are numerically investigated. The pylon induced drag is decomposed by the TDB method. Comparisons of the TFN configuration and PN configuration show that the jet effect will decrease the lift and pitching moment of the PN configuration. A negative pressure region on the pylon is also induced. These effects, as well as the difference between the TFN drag and the PN's aerodynamic drag plus thrust loss, need to be noticed by the aircraft designer.

ACKNOWLEDGEMENTS

This work was supported by the National Key Basic Research Program of China (No.2014CB744801), the National Natural Science Foundation of China (Nos. 11102098 and11372160) and the Aeronautical Science Foundation of China (No. 2013ZA58002).

REFERENCES

1. Laban M. Aircraft drag and thrust analysis. Amsterdam: National Aerospace Laboratory; 2000. Report No.: NLR-TP-2000-473.
2. Tan ZG, Chen YC, Li J, Zhang M. Numerical simulation method for the powered effects in airframe/propulsion integration analysis. J Aerospace Power 2009;24(8):1766–72 Chinese.
3. Rooney EC. Thrust and drag: its prediction and verification. In: Covert EE, editor. Thrust-drag accounting methodology. Progress in astronautics and aeronautics. Reston: AIAA; 1985. p. 29–45.
4. Zheng WL, Wang YK, Shan JX. Interference mechanism of engine exhaust on civil aircraft drag performance. J Aircraft 2012;49(6):2001–6.
5. Hunt BL, Gowadia NS. Determination of thrust and throttledependent drag for fighter aircraft. Reston: AIAA; 1981 Report No.: AIAA-1981-1692.
6. Kooi JW, Haij L, Hegen GH. Engine simulation with turbofan powered simulators in the German-Dutch wind tunnels. Reston: AIAA; 2002 Report No.: AIAA-2002-2919.

7. Bousquet JM. Survey of engine integration testing in ONERA wind tunnels. Reston: AIAA; 2005 Report No.: AIAA-2005-3705.

8. Decher R, Tegeler DC. High accuracy force accounting procedures for turbo powered simulator testing. Reston: AIAA; 1975 Report No.: AIAA-1975-1324.

9. Dusa D, Lahti DJ, Berry D. Investigation of subsonic nacelle performance improvement concept. Reston: AIAA; 1982 Report No.: AIAA-1982-1042.

10. MIDAP Study Group. Guide to in-flight thrust measurement of turbojets and fan engines. Paris: Advisory Group for Aerospace Research and Development; 1979 Report No.: AGARD-AG-237.

11. von Geyr HF, Rossow CC. A correct thrust determination method for turbine powered simulators in-wind tunnel testing. Reston: AIAA; 2005 Report No.: AIAA-2005-3707.

12. Wright FL. Comparison of least squares curve fit and individual sample statistical analysis results of calibration data for the velocity coefficient of a flow nacelle. Reston: AIAA; 1994 Report No.: AIAA-1994-2587.

13. Hunt DN. Experimental techniques used to evaluate propulsion system interference effects on the cruise configuration of the Boeing C-14. Reston: AIAA; 1979 Report No.: AIAA-1979-0335.

14. Malecki RE, Lord WK. Aerodynamic performance of exhaust nozzles derived from CFD simulation. Reston: AIAA; 1995 Report No.: AIAA-1995-2623.

15. Chen AW, Tinoco EN. PAN AIR application to aero-propulsion integration. J Aircraft 1984;21(3):161–7.

16. Nesbitt E, Mengle V, Czech M. Flight test results for uniquely tailored propulsion-airframe aeroacoustic chevrons: community noise. Reston: AIAA; 2006 Report No.: AIAA-2006-2438.

17. Frate FC, Khavaran A. An aerodynamic and acoustic assessment of convergent-divergent nozzles with chevrons. Reston: AIAA; 2011 Report No.: AIAA-2011-0976.

18. Barberie FJ, Wick AT, Hooker JR. Low speed powered lift testing of a transonic cruise efficient STOL military transport. Reston: AIAA; 2013 Report No.: AIAA-2013-1099.

19. Abdol-Hamid KS. Commercial turbofan engine exhaust nozzle flow analyses. J Propul Power 1993;9(3):431–6.

20. Winkler CM, Dorgan AJ. BCFD analysis for the 1st AIAA Propulsion Workshop: nozzle results. Reston: AIAA; 2013 Report No.: AIAA-2013-3731.

21. Zhang YF, Chen HX, Zhang M. Performance prediction of conical nozzle using Navier-Stokes computation. J Propul Power 2015;31(1):192–203.

22. Spotts N, Guzik S, Gao XF. A CFD analysis of compressible flow through convergent-conical nozzles. Reston: AIAA; 2013 Report No.: AIAA-2013-3734.

23. Chen SY, Chen YC, Xia ZH. Constrained large-eddy simulation and detached eddy simulation of flow past a commercial aircraft at 14 degrees angle of attack. Sci China Phys Mech Astron 2013;56(2):270–6.

24. Zhang YF, Chen HX, Fu S. Improvement to patched grid with high-order conservative remapping method. J Aircraft 2011;48(3): 884–93.

25. Zhang YF, Chen HX, Fu S. A Karman-vortex generator for passive separation control in a conical diffuser. Sci China Phys Mech Astron 2012;55(5):828–36.

26. Roe PL. Approximate Riemann solvers, parameter vectors, and difference schemes. J Comput Phys 1981;43(2):357–72.

27. Menter FR, KuntzM, Langtry R. Ten years of industrial experience with the SST turbulence model. In: Hanjalic K, Nagano Y, Tummers M, editors. Proceeding of the 4th international symposium on turbulence, heat and mass transfer. 2003. p. 625–32.

28. Thornock RL, Brown EF. An experimental study of compressible flow through convergent conical nozzles, including a comparison with theoretical results. J Basic Eng 1972;94(4):926–30.

29. Fu DB, Yu Y, Niu QL. Simulation of underexpanded supersonic jet flows with chemical reactions. Chin J Aeronaut 2014;27(3): 505–13.

30. Hunter CA, Thomas RH, Abdol-Hamid KS. Computational analysis of the flow and acoustic effects of jet-pylon interaction. Reston: AIAA; 2005 Report No.: AIAA-2005-3083.

31. Massey SJ, Elmiligui AA, Hunter CA. Computational analysis of a chevron nozzle uniquely tailored for propulsion airframe aeroacoustics. Reston: AIAA; 2006 Report No.: AIAA-2006-2436.

32. Vassberg JC, Tinoco EN, Mani M. Summary of the 4th AIAA Computational Fluid Dynamics Drag Prediction Workshop. J Aircraft 2014;51(4):1070–89.

33. Levy DW, Laflin KR, Tinoco EN. Summary of data from the Fifth AIAA CFD Drag Prediction Workshop. Reston: AIAA; 2013 Report No.: AIAA-2013-0046.

34. Li J, Gao ZH, Huang JT, Zhao K. Aerodynamic design optimization of nacelle/pylon position on an aircraft. Chin J Aeronaut 2013;26(4):850–7.

CITATION

Yufei Zhang, Haixin Chen, Song Fu, Miao Zhang, Meihong Zhang, Drag prediction method of powered-on civil aircraft based on thrust drag bookkeeping, Chinese Journal of Aeronautics, Volume 28, Issue 4, August 2015, Pages 1023-1033, ISSN 1000-9361, http://dx.doi.org/10.1016/j.cja.2015.06.015.

CHAPTER 7

Multi-Body Dynamic System Simulation of Carrier-Based Aircraft Ski-Jump Takeoff

Wang Yangang , Wang Weijun , Qu Xiangju

School of Aeronautical Science and Engineering, Beihang University, Beijing 100191, China

ABSTRACT

The flight safety is threatened by the special flight conditions and the low speed of carrier-based aircraft ski-jump takeoff. The aircraft carrier motion, aircraft dynamics, landing gears and wind field of sea state are comprehensively considered to dispose this multidiscipline intersection problem. According to the particular naval operating environment of the carrier-based aircraft ski-jump takeoff, the integrated dynamic simulation models of multi-body system are developed, which involves the movement entities of the carrier, the aircraft and the landing gears, and involves takeoff instruction, control system and the deck wind disturbance. Based on Matlab/Simulink environment, the multi-body system simulation is realized. The validity of the model and the rationality of the result are verified by an example simulation of carrier-based aircraft ski-jump takeoff. The simulation model and the software are suitable for the study of the multidiscipline intersection problems which are involved in the performance, flight quality and safety of carrier-based aircraft takeoff, the effects of landing gear loads, parameters of carrier deck, etc.

INTRODUCTION

Ski-jump takeoff is one of the main takeoff modes for a carrier-based aircraft. Compared with catapult launch, it takes longer time on deck-running, acquires lower speed at ramp exit and greater effects by the aircraft carrier motion, wind field disturbance and launching time. Consequently it is important to take account of those factors comprehensively. From the viewpoint of system engineering, it is necessary to build an integrated system simulation model considering all the kinds of important influencing factors. This model brings about theoretical and practical significance not only to the research of flight dynamic problem referring to carrier-based aircraft, but also to the analysis of multidiscipline intersection problems including the suitability of carrier and aircraft, the motion coupling of deck, aircraft body and landing gears, as well as the safety of carrier-based aircraft takeoff or landing, etc.

Many studies on the simulation of ski-jump takeoff process have been developed recently, including modeling of carrier-based aircraft motion, modeling of aircraft carrier motion and modeling of wind field disturbances, etc.[1, 2, 3, 4 and 5] Particularity and complexity of the physical system make the research difficult on system modeling and simulating.[6 and 7] Much work about modeling of the complete system needs to be carried out so far. Based on the latest researches of simulation modeling of ski-jump takeoff, this paper has built an integrated system simulation model taking account of multi-motion-body coupling between carrier, aircraft and landing gears, as well as the wind field induced by the aircraft carrier, the command decision on deck, and the control policy of pilot.

PHYSICAL SYSTEM ANALYSIS

The flight path trajectory from a ski-jump takeoff can be divided into two phases: deck run and part-ballistic flight, as shown in Fig. 1. The deck run is from releasing the brakes up to the ramp exit; and the part-ballistic flight is from the moment of leaving board to the fully wing-borne flight. The closed force vector pentagons depict the development of the aircraft acceleration denoted by a hollow arrow during the ski-jump takeoff, where W represents the weight, L the lift, D the drag and T the thrust.

Aircraft carrier motion

Aircraft carrier is a moving platform with pitch, roll and heave motions. This will alter attitude angles and flight-path angles of the carrier-based aircraft at the ramp exit stochastically. At the same time, the moving deck will induce the transport acceleration, which affects the flight state of leaving.

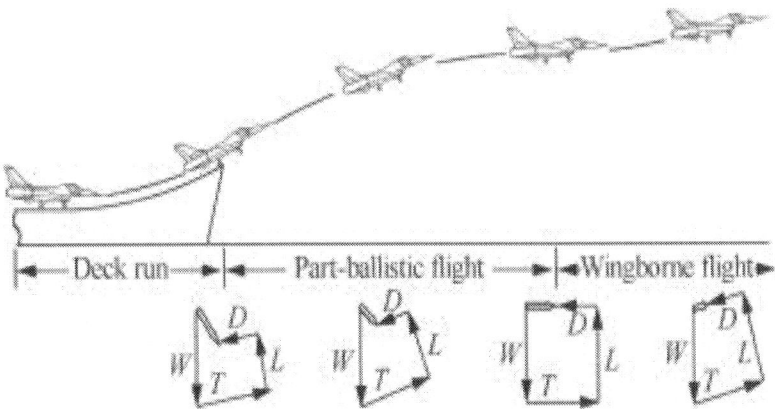

Figure 1. Schematic of carrier-based aircraft ski-jump takeoff.

Bow gust and ground effect

The speed of ski-jump takeoff is much lower than that of catapult launch. When the carrier-based aircraft leaves the board, the ground effect vanishes and the lift is not enough to balance the gravity. The aircraft maintains the early climb away from the aircraft carrier with the help of the upward path angle established by the inclination angle of the deck. Meanwhile, the angle of attack (AOA) increases quickly in short time because of the flow over the carrier bow, especially the sudden upward gust fore of the ramp. It is dangerous for stalling and influences the safety of takeoff. For the vanishing of the ground effect when the aircraft flies away the carrier deck, the part of the lift loses suddenly. Then the flight track will sink. This is another factor influencing the safety of takeoff.

Dynamics of landing gears

The carrier-based aircraft starts the takeoff roll on the deck when the wheel chock is laid down. At this moment, the front landing gear dumper begins to stretch from the compressed state caused by the full reheating condition of the aircraft. And the main landing gears are further compressed because of bearing more weight. Along with the dumpers of main landing gears compressed excessively, the gas pressure rises to make the compressing stop and then the dumpers stretch. For this repetitive process, the relative pitch angle of the carrier-based aircraft to the aircraft carrier continuously rises and falls in the taxiing stage.

After the carrier-based aircraft rolls on the ramp, the landing gears are shocked respectively for the changed curvature of the deck. The dumpers of landing gears are compressed and stretched again to induce a new wave of the relative pitch angle from its leveling out. This oscillation will change the attitude angle and the angle of attack of carrier-based aircraft at the ramp exit. And the takeoff condition will be affected.

Multi-kinetic-bodies coupling

The deformation of the tires and dumpers of landing gears will change the forces acting on the bodies connected to it. So the takeoff characteristic is affected by the landing gear system significantly. There are serious couplings between the landing gears and the aircraft body.

Launching time decision

Carrier-based aircraft launching is a multiplayer and multi-machine system dynamic process. Besides, it is not only affected by the carrier motion and the disturbance of special wind environment, but also involves the collaborative decision control among the launching signal officer (LSO) and the pilot.

The complex environment in takeoff process, the coupling of multi-body motion and the multiplayer collaborative decision control are all the influencing factors of carrier-based aircraft safety.

BUILDING METHOD OF SUBSYSTEM MODELS

Modeling of aircraft carrier motion

In engineering practice, the aircraft carrier motion under the action of sea waves widely described by statistical analysis technique is usually regarded as an ergodic stochastic process. Stationary stochastic process theory is used, supposing the spectrum function of aircraft carrier motion is static continuous, time invariant and zero mean.[8]

Modeling of aircraft carrier disturbances

According to the physical characteristics and causes of the flow around aircraft carrier, the spatial distribution of steady components, the engineering calculation methods of free turbulent components, random turbulent components and periodic components are given in American standard MIL-1797A.[9] Certain experience about the simulation of the flow around aircraft carrier has been gained at present.[10]

Ground effect influences

The theoretical formula and modifier formula for calculating the aerodynamic data affected by ground effect have been built in engineering practice. The changes of aerodynamic forces caused by ground effect can be calculated in various flight heights from ground.[11]

Dynamic modeling of flexible multi-body system

The whole multi-body dynamic system consists of aircraft carrier, carrier-based aircraft body, moveable parts of three landing gears whose two ends connect with nonholonomic constraints (the displacement constraint between tire and deck) and holonomic constraints (the dumper treated as prismatic joint) respectively. The schematic diagram of the multi-body system is shown in Fig. 2.[12] The supporting force N_i ($i = 1$, 2, 3 represent front tire, right main tire and left main tire individually) and the friction f_i are included in the derivation all the time, no matter whether the tires touch the deck or not. Only the existence of them needs to be defined according to the positions of the tires (contact the deck or not). Where D_i is the contact point between the deck and the tire. G_i represents the centre of gravity of the movable part of the landing gears, where i ($i = 1$, 2, 3) indicates the nose gear, the right and left main gear respectively. The aircraft-body coordinate system $Bx_by_bz_b$ is fixed to the airframe where the origin B is placed at the aircraft mass center. V represents the velocity vector of the aircraft, α the AOA and α_t the angle between the thrust T and x_b.

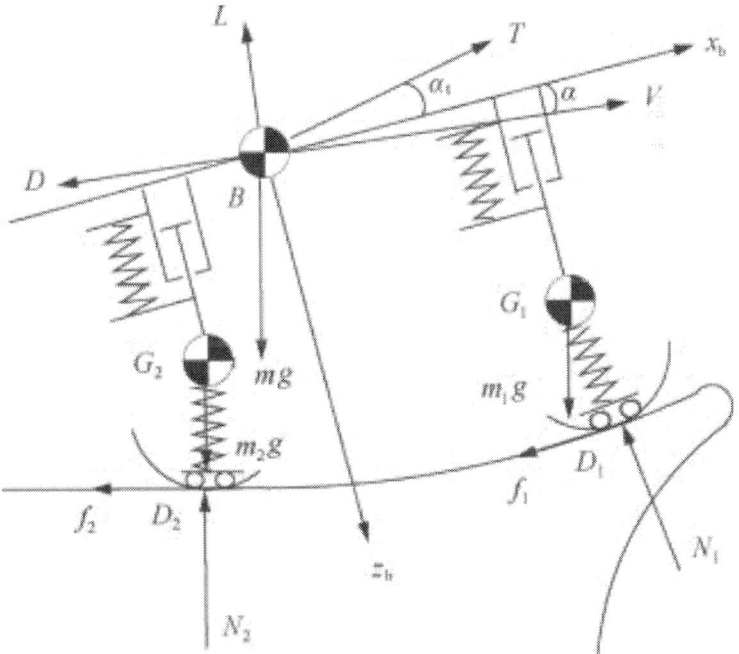

Figure 2. Simplification of multi-body dynamic system.

The effects of a moving carrier-based aircraft on an aircraft carrier motion are negligible as the mass of the aircraft is nearly three orders of magnitude less than the aircraft carrier. Therefore the carrier motion is independent of the carrier-based aircraft and regarded as an input of the multi-body dynamic system (MBDS). Similarly, the influence of the relative movement of the strut dampers of landing gears on the position of the c.m. (center of mass) of the aircraft relative to the airframe can be ignored, since the mass of the movable parts of landing gears is much less than that of the aircraft (about 2.2% of the total mass of the aircraft). So it is assumed that the position of the aircraft c.m. relative to the airframe remains unchanged, and the mass centers of moveable parts of landing gears are located at their respective wheel axles.

Flight instruction and control module

The LSO is responsible for the safety of the carrier-based aircraft takeoff. Before the deck run, the aircraft is attached to the flight deck by the holdback fitting to enable the engine to run up to full power. After the pilot signals the LSO that it is ready, the commander will make a right judgment by considering carrier motion, aircraft characteristics and flight mission, etc. If the takeoff decision is made, the LSO will give signals immediately to the launch operator to release the wheel gear, and the carrier-based aircraft will then start rolling and complete the takeoff process. Otherwise a right time shall be waited for. The time decision-making system for carrier-based aircraft launching is shown as Fig. 3.[13]

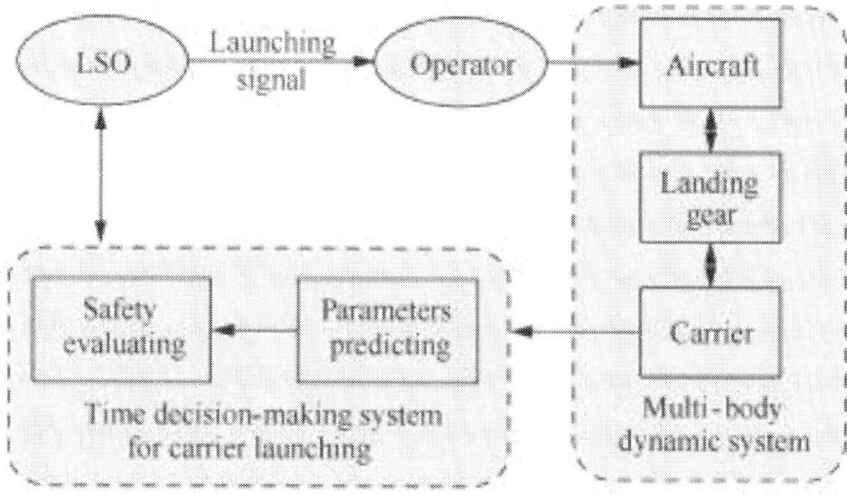

Figure 3. Time decision-making system for carrier-based aircraft launching.

DYNAMIC MODELING OF MULTI-BODY SYSTEM

The dynamic equations of the aircraft and landing gears are built with reference to the simplified mechanism of the multi kinetic bodies shown in Fig. 2. The tensor symbols are used to make the model form of the multi-body dynamic system simple and axiomatic.

Dynamic model of mass center of aircraft

The acceleration of the carrier-based aircraft c.m. with respect to (wrt) the aircraft carrier frame is given by Newton's second law:

$$
\left[\frac{d^2 s_{BS}}{dt^2}\right]^S = \frac{[T]^{SW}[F_a]^W + [T]^{SB}[F_t]^B + \sum_{i=1}^{3}[T]^{SD_i}[F_{di}]^{D_i}}{m_B + m_{G_1} + 2m_{G_2}} + [T]^{SI}[g]^I - \underbrace{\left[\frac{d\Omega^{SI}}{dt}\right]^S[s_{BS}]^S}_{\text{Tangential acceleration}} - \underbrace{2[\Omega^{SI}]^S\left[\frac{ds_{BS}}{dt}\right]^S}_{\text{Coriolis acceleration}} - \underbrace{[\Omega^{SI}]^S[\Omega^{SI}]^S[s_{BS}]^S}_{\text{Centripetal acceleration}} - \underbrace{[T]^{SI}\left[\frac{d^2 s_{SI}}{dt^2}\right]^I}_{\substack{\text{Inertia acceleration of the aircraft} \\ \text{carrier frame}}}
$$

(1)

where the superscript S means the aircraft carrier coordinate system, B the aircraft body frame, W the wind frame, I the inertial frame, and D_i the deck-tire contact point frame. The subscript I indicates the original point of the inertial frame, B the c.m. of the aircraft body, S the c.m. of the carrier body, and G_i the c.m. of the movable part of the landing gears. The deck-tire contact point frame is as follows: the contact point D_i between the deck and the i th tire is the origin, the plane $D_i x_{di} y_{di}$ is the tangent plane of the deck through the point D_i, the axis x_{di} points to the bow, while the axis y_{di} points to the right, and the axis z_{di} points downwards, perpendicular to the plane $D_i x_{di} y_{di}$. So $[T] SD_i$ indicates the transfer matrix from the deck-tire contact point frame to the aircraft carrier frame. $[T]^{SW}$, $[T]^{SB}$ and $[T]^{SI}$ indicate the transfer matrix from wind frame to the carrier frame, the matrix from aircraft body frame to the carrier frame, and the matrix from inertial frame to the carrier frame respectively. m_B is the mass of the aircraft body and m_{Gi} the mass of the movable parts of the landing gear. $[g]^I$ is the gravity acceleration expressed in the inertial frame. The displacement vector $[s_{BS}]^S$ of point B wrt point S, and $[s_{SI}]^S$ of point S wrt point I, are both expressed in the aircraft carrier frame. $[F_a]^W$ indicate the aerodynamics force expressed in the wind frame, and $[F_t]^B$ thrust force expressed in the aircraft body frame. $[F_{di}]^{D_i}$ expressed in the contact point frame, means the deck reaction acting on the tire i (i = 1, 2, 3 represent front tire, right main tire and left main tire individually). $[\Omega^{SI}]^S$ is the skew symmetric form of $[\omega^{SI}]^S$, the angular velocity of the aircraft carrier wrt the inertial frame, expressed in the carrier frame.

Rotation dynamic model of aircraft body

Using the angular momentum theorem, the angular acceleration of the aircraft wrt inertial frame described in the aircraft body frame is as follows:

$$\left[\frac{d\omega^{BI}}{dt}\right]^{B} = \{[I_{B}^{B}]^{B}\}^{-1}\left\{[M_{a}]^{B} + \sum_{i=1}^{3}[S_{D_{i}B}]^{B}[T]^{BD_{i}}[F_{di}]^{D_{i}} - \left[\frac{dI_{B}^{B}}{dt}\right]^{B}[\omega^{BI}]^{B} - [\Omega^{BI}]^{B}[I_{B}^{B}]^{B}[\omega^{BI}]^{B}\right\}$$

(2)

where M_{a} is the aerodynamic moment, I_{B}^{B} the moment of inertia wrt the mass center B. $[\Omega^{BI}]^{B}$ is the skew symmetric form of $[\omega^{BI}]^{B}$, the angular velocity of the aircraft wrt the inertial frame, expressed in the body frame B . $[S_{DiB}]^{B}$ is the skew symmetric form of $[s_{DiB}]^{B}$, the displacement vector of the contact point D_i wrt point B , expressed in the aircraft body frame. $[T]BD^{i}$ indicates the transfer matrix from the D_i frame to the B frame. The effects of spinning rotors are negligible.

Dynamic model of landing gears

Using Newton's law (e.g., the front landing gear, G_1 is the mass center):

$$\left[\frac{d^{2}s_{G_{1}B}}{dt^{2}}\right]^{B} = \frac{[T]^{BS}[T]^{SD_{1}}[F_{d1}]^{D_{1}} + [F_{b1}]^{B} + [F_{l1}]^{B}}{m_{G_{1}}} + [g]^{B} - \underbrace{\left[\frac{d\Omega^{BI}}{dt}\right]^{B}[s_{G_{1}B}]^{B}}_{\text{Tangential acceleration}} - \underbrace{2[\Omega^{BI}]^{B}\frac{ds_{G_{1}B}}{dt}^{:B}}_{\text{Coriolis acceleration}} - \underbrace{[\Omega^{BI}]^{B}[\Omega^{BI}]^{B}[s_{G_{1}B}]^{B}}_{\text{Centripetal acceleration}} - \underbrace{[T]^{BI}\frac{d^{2}s_{BI}}{dt^{2}}^{:I}}_{\substack{\text{Inertia acceleration of the aircraft}\\\text{expressed in the body frame}}}$$

(3)

where the displacement vector $[s_{G1B}]^{B}$ of point G_1 wrt point B, the dumper forces (air pressure in air cavity, friction and damping force of the oil) $[F_{b1}]^{B}$ and the limit force (caused by the piston displacement constraint) $[F_{l1}]^{B}$ are all expressed in the B frame. $[g]^{B}$ is the gravity acceleration expressed in the B frame.

SYNTHESIS OF SIMULATION SYSTEM MODEL

The implementations of the system model can be divided into five modules shown in Fig. 4. The module of kinematic and dynamic system of the aircraft and the landing gears, which generates all the necessary state variables by solving the dynamic equations, is the primary module of the system. The module of environmental factors defines the carrier motions, deck geometry, carrier disturbances and ground effects. The time decision-making system module predicts the decision-making parameters by analyzing carrier motions sampling, and then makes safety evaluation whether to launch or not. The initial states solving module determines the initial condition of simulation. The data record and plot module function in data recording and processing.

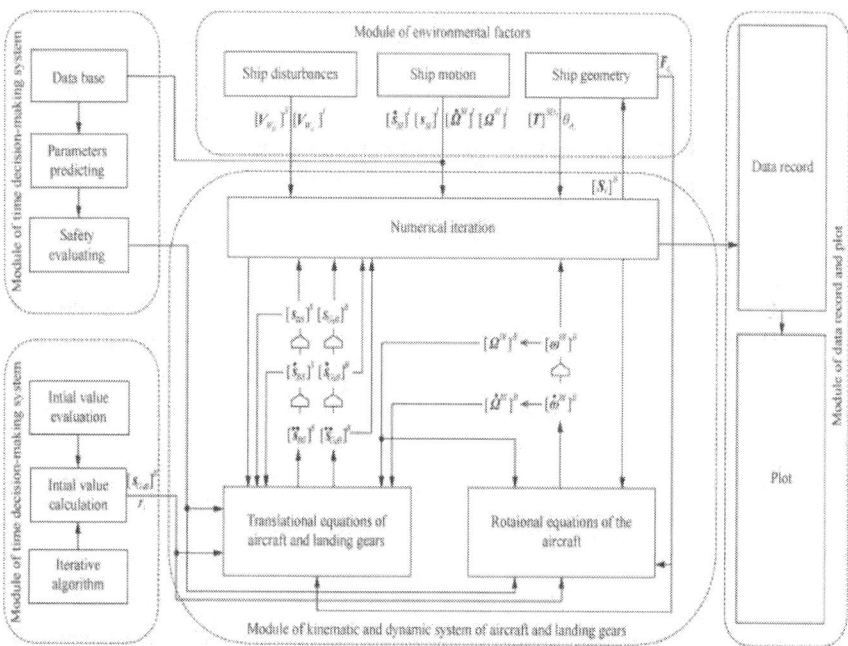

Figure 4. Synthesis of the MBDS.

Where $[\dot{s}_{BS}]^S$ and $[\ddot{s}_{BS}]^S$, expressed in the S frame, are the first and second rotational derivative of the displacement vector $[s_{BS}]^S$. $[\dot{s}_{G_iB}]^B$ and $[\ddot{s}_{G_iB}]^B$, expressed in the B frame, are the first and second rotational derivative of the displacement vector $[s_{GiB}]^B$. $[\dot{s}_{SI}]^I$ expressed in the I frame, is the first rotational derivative of the displacement vector $[s_{SI}]^I$. $[\dot{\Omega}^{BI}]^B$ is the skew symmetric form of $[\dot{\omega}^{BI}]^B$, the first rotational derivative of $[\omega^{BI}]^B$, expressed in the B frame. $[\dot{\Omega}^{SI}]^I$ is the skew symmetric form of $[\dot{\omega}^{SI}]^I$, the first rotational derivative of $[\omega^{SI}]^I$, expressed in the S frame. $[V_{WD}]^S$ and $[V_{WA}]^I$, expressed in the S and I frame respectively, are the wind velocity caused by the carrier deck disturbance and free atmosphere disturbance. θ_{di} is the angle between x_S of the aircraft carrier frame and the local deck plane where D_i lies. r_i is the radius of the tire i.

SIMULATION EXAMPLE

Dynamic characteristics of multi-body system

Based on the model developed above, a simulation for carrier-based aircraft ski-jump takeoff is carried out in Sea State 4. The flight deck of Varyag is used, of which the ramp length is 55 m. The aircraft carrier speed is 12.86 m/s. The typical dynamic characteristics of the aircraft carrier, the carrier-based aircraft and the landing gears are shown in Fig. 5, Fig. 6, Fig. 7, Fig. 8, Fig. 9, Fig. 10, Fig. 11, Fig. 12 and Fig. 13.

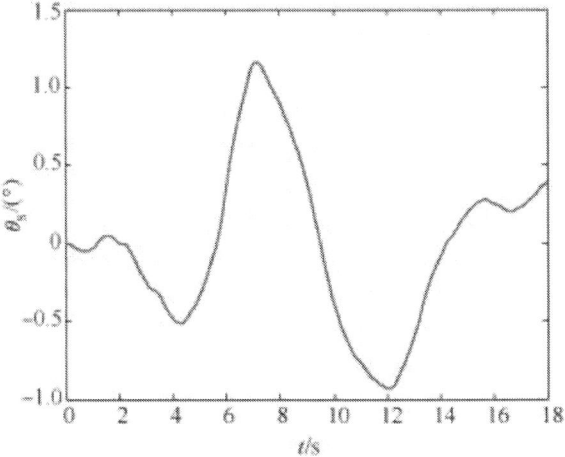

Figure 5. Time history of the aircraft carrier pitch angle.

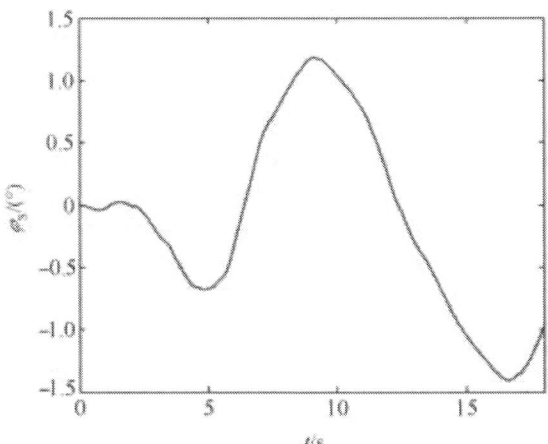

Figure 6. Time history of the aircraft carrier roll angle.

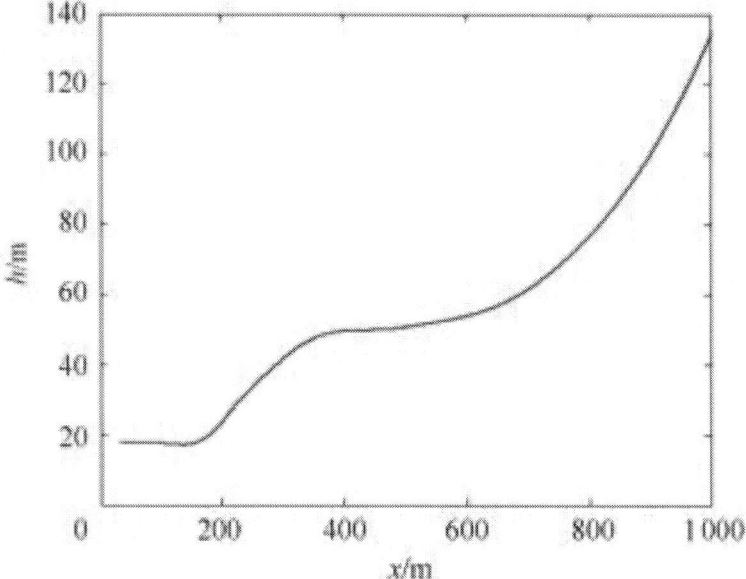

Figure 7. Flight trajectory profile in ski-jump takeoff.

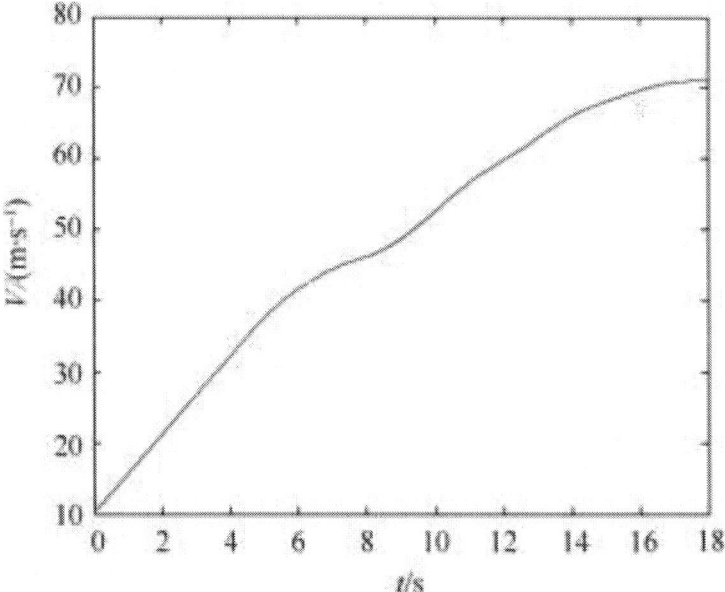

Figure 8. Time history of the velocity.

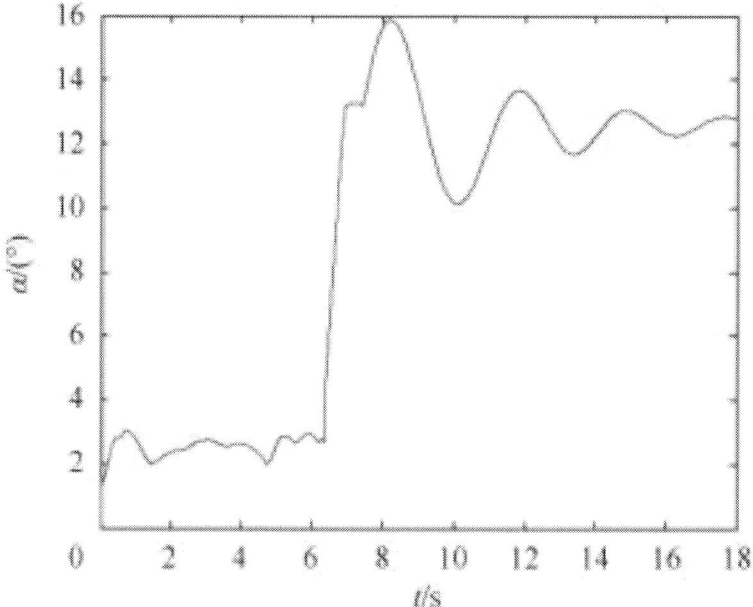

Figure 9. Time history of the angle of attack AOA.

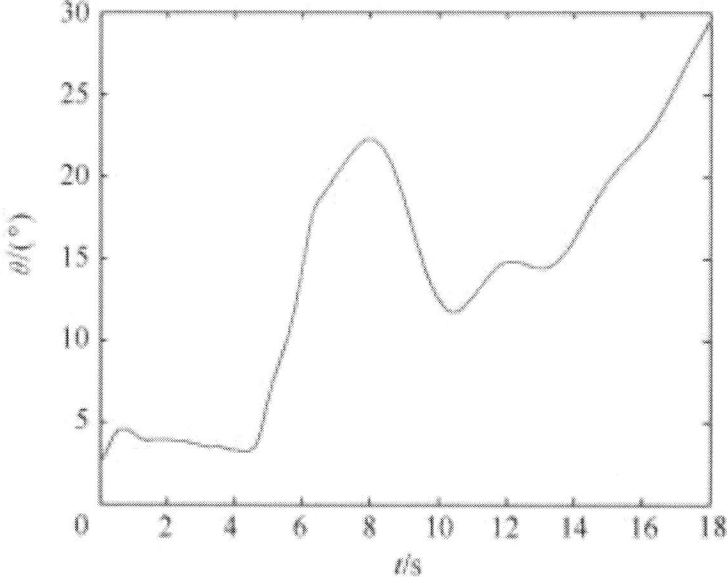

Figure 10. Time history of the pitch angle.

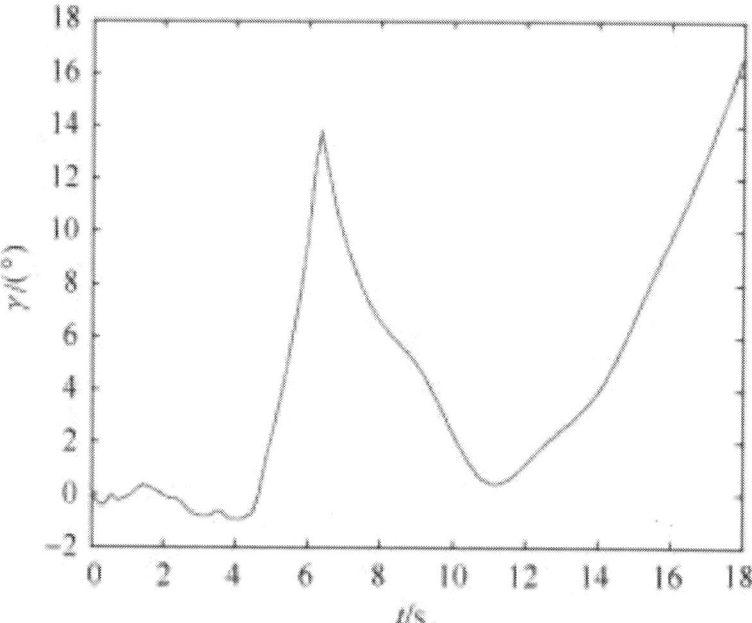

Figure 11. Time history of the flight-path angle.

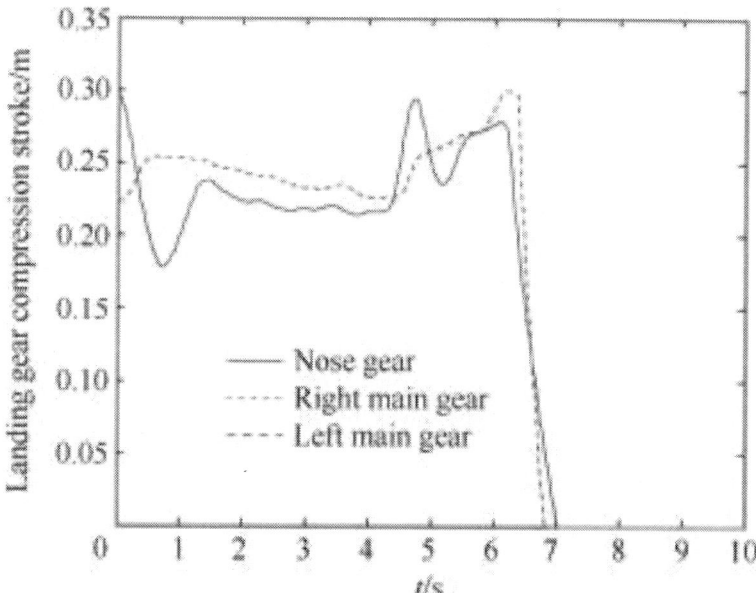

Figure 12. Stroke-time histories of landing gears.

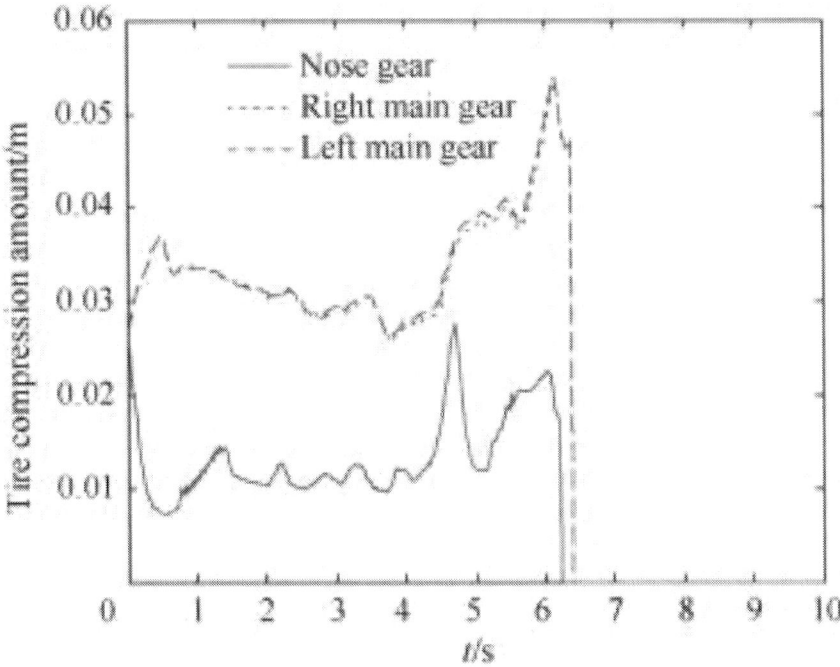

Figure 13. Compression-time histories of tires.

The carrier-based aircraft starts to run on the flat deck at 0 s, arrives at ramp entry at 6.7 s, and leaves off the deck at 8.2 s. The influences on the performance and handling quality of the takeoff by the aircraft carrier motion and wind disturbance are reflected in main flight parameters and landing gears responses, which can be used to study flight safety of the takeoff. So the simulation system model has practical significance for the design of aircraft carrier deck, landing gears, aircraft aerodynamic configuration and structure, and for the analysis of the flight dynamics and flight safety.

Influences of takeoff time and pilot control

The simulations are carried out in various deck motion states and various pilot controls when aircraft leaving aircraft carrier. The takeoff time is determined by the deck commander. Then the pilot responds to the commander's instruction to start taking off. The takeoff time decides if the aircraft carrier noses up or down at the moment when aircraft leaves the deck, see Table 1. The attitude of the aircraft carrier can affect the takeoff performance and safety of the carrier-based aircraft significantly. The aerodynamic drag can be reduced when the aircraft is taxiing on the carrier with an angle of attack close to zero. The proper angle of attack will be determined when leaving by the curved deck rotating the aircraft and pilot pulling stick. Analysis of the takeoff time and the pilot control helps to study the influences of human instruction and control on takeoff safety.

Table 1. Simulation settings for ramp exit.

Control mode	Ship motion	Time for pulling stick
B−	Bow up	At ramp exit
B+	Bow up	1 s after ramp exit
C−	Bow down	At ramp exit
C+	Bow down	1 s after ramp exit

The results are shown in Fig. 14, Fig. 15, Fig. 16, Fig. 17 and Fig. 18.

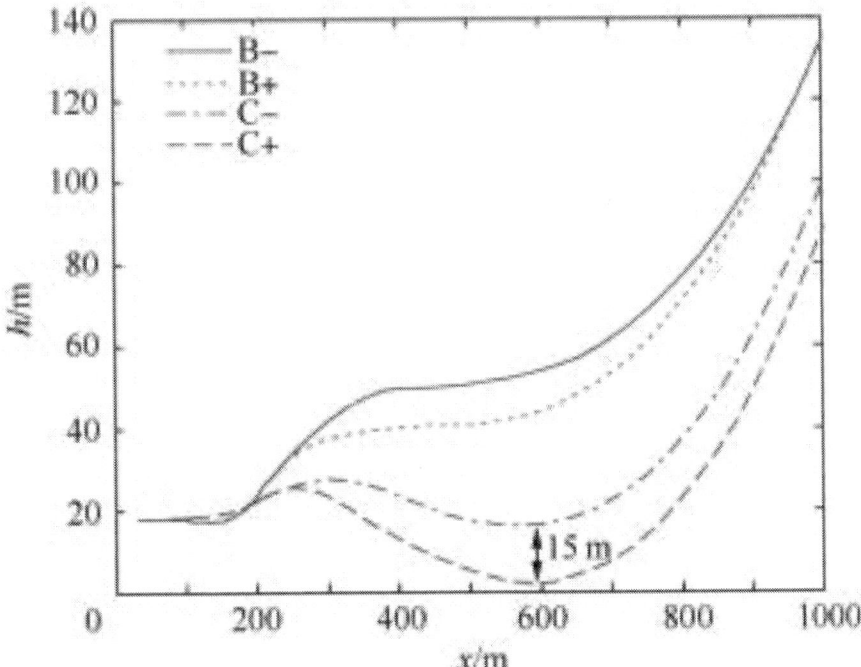

Figure 14. Comparison of ski-jump takeoff trajectories in different control modes.

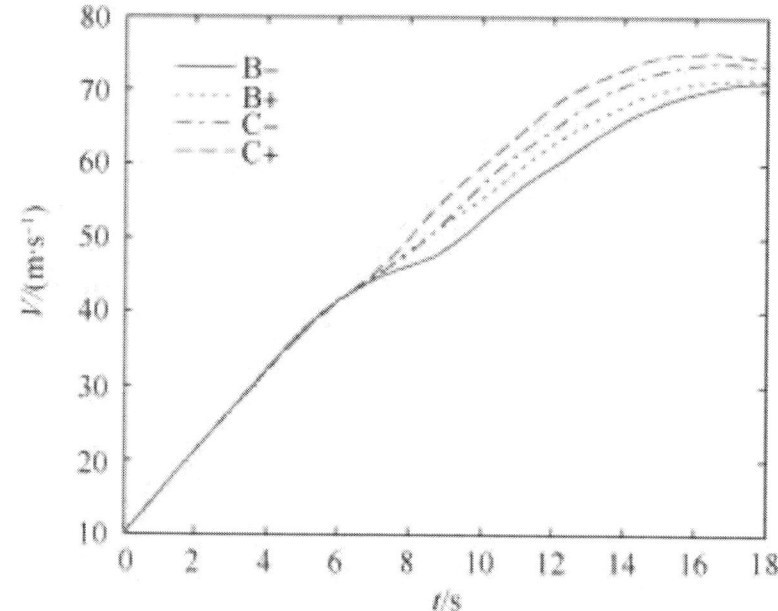

Figure 15. Comparison of airspeeds in different control modes.

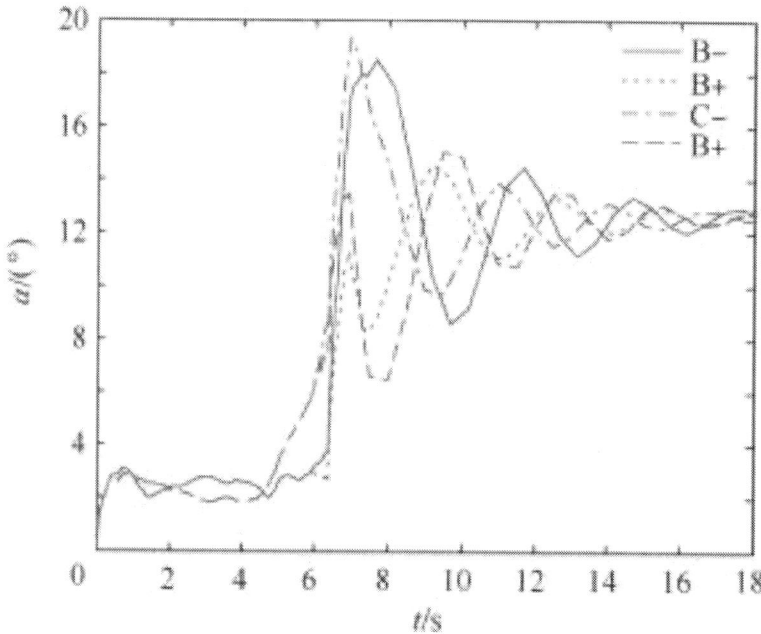

Figure 16. Comparison of angles of attack in different control modes.

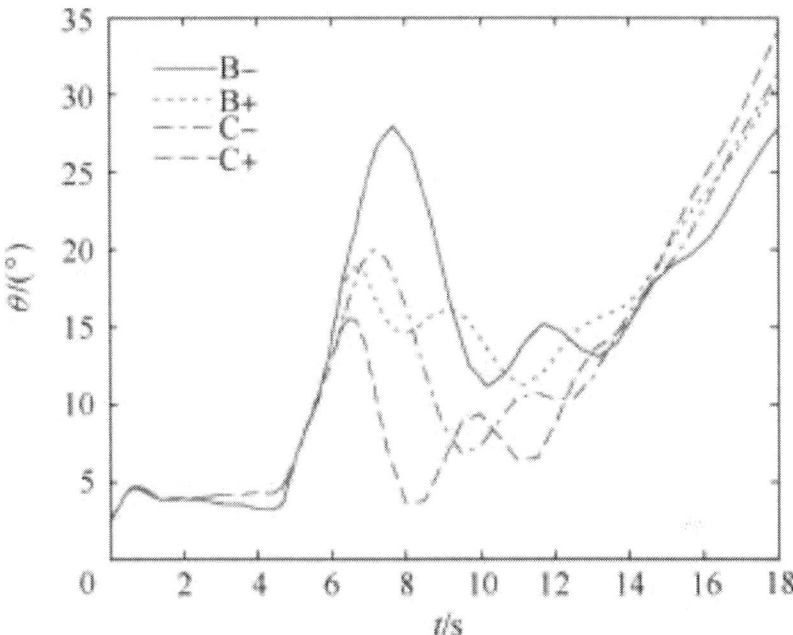

Figure 17. Comparison of pitch angles in different control modes.

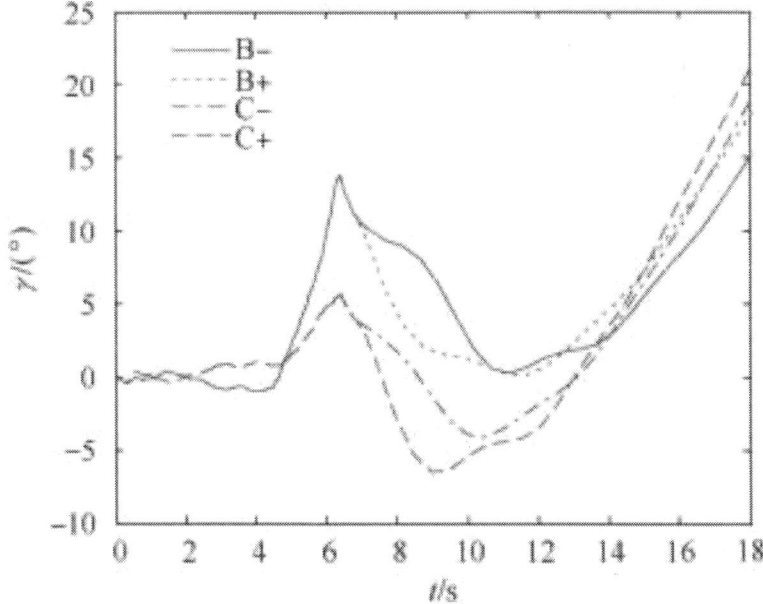

Figure 18. Comparison of flight path angles in different control modes.

The figures show that the in-time pulling stick will be helpful to improve the aircraft flight trajectory after leaving the aircraft carrier, but it goes against to restrain the angle of attack increasing excessively. Delaying control will go opposite.

If pilot pulls up earlier, the aircraft will pitch up earlier. Then for the larger pitch angle, the thrust will be distributed more on the vertical direction to make up for the deficiency of the aircraft lift and to improve the flight track. For example, when the aircraft carrier is pitching down (state C), earlier control will upraise the trajectory about 15 m at the lowest point. When the carrier is pitching up (state B), the initial trajectory will be raised obviously. For the lack of the aircraft lift, the flight path angle will decrease rapidly after leaving the aircraft carrier. Therefore, if the pitch angle is large prematurely, the angle of attack will increase rapidly and the aircraft will be dangerous to stall. The existence of the bow gust will aggravate the situation.

If the pilot control is delayed, the excessive growth of the angle of attack and stall will be avoided. In the two-wheels-taxiing stage, the support reaction produces nose-down moment with the front landing gear free; larger angle of attack can produce nose-down moment too. So delaying control until the angle of attack decreases and the speed increases to a value large enough will make the aircraft not easy to stall. However, if waiting for a long time to control, the path angle will decrease quickly for the lack of lift in the waiting time. It will miss the opportune time to upraise the trajectory despite the forthcoming control. The aircraft sinks too much, which would threaten the takeoff safety. Therefore, the control time should be selected properly.

CONCLUSIONS

The simulation modeling of carrier-based aircraft ski-jump takeoff is complicated. This paper builds the relatively complete system model of carrier-based aircraft ski-jump takeoff to resolve the problems of the coupling among multi-motion bodies and flight environment, as well as the problems of the cooperative instructions control. This system model takes into account three main effects: the coupling of carrier, aircraft body and the landing gears; the influences on the carrier motion by sea state and on the flight by the induced wind field; the influences on the aircraft flight by the cooperative instructions control among deck commanders and pilot. Two simulation examples show that the system model can describe the dynamic characteristics of all the movement bodies reasonably. It has practical significance for the multidisciplinary intersect problem in the design of carrier deck, design of landing gears and aircraft body. This system model can be used to analyze the influencing factors of flight safety comprehensively, such as flight environment, human decision-making control, etc., which is supposed to play an important role in flight training.

REFERENCES

1. Jin YJ, Wang LX. The process and mathematic description of skijump take off of shipboard aircraft. Flight Dyn 1994;12(3):45–52 in Chinese.
2. Peng J, Jin CJ. Research on the numerical simulation of aircraft carrier air wake. J Beijing Univ Aeronaut Astronaut 2000;26(3):340–3 in Chinese.
3. Wang WJ, Qu XJ, Guo LL. Multi-agent based hierarchy simulation models of carrier-based Aircraft catapult launch. Chin J Aeronaut 2008;21(3):223–31.
4. Wang WJ, Guo LL, Qu XJ. Analysis of the mechanics for skijump takeoff. J Beijing Univ Aeronaut Astronaut 2008;34(08):887–90 in Chinese.
5. Lawrence JT. Milestones and developments in US naval carrier aviation – part II. AIAA-2005-6120; 2005.
6. Zhang W, Zhang Z, Zhu QD, et al. Dynamics model of carrierbased aircraft landing gears landed on dynamic deck. Chin J Aeronaut 2009;22(4):371–9.
7. Fry A, Cook R, Revill N. CVF ski-jump ramp profile optimisation for F-35B. Aeronaut J 2009;113(1140):79–85.
8. Durand TS, Teper GL. An analysis of terminal flight path control in carrier landing. AD606040; 1964.
9. Military Standard MIL-STD-1797A. Military standard flying qualities of piloted aircraft. VA: Defense Quality and Standardization Office 1990.
10. Deveson KH. STOVL Carrier operations – comparison of safe launch criteria and MTOW sensitivities using APOSTL. AIAA- 1997-5516; 1997.
11. Zhang NP. Ground effect on the take-off characteristics of seabased aircraft. Acta Aerodyn Sin 1992;10(4):451–6 in Chinese.
12. Liu WW, Qu XJ. Modeling of carrier-based aircraft ski jump takeoff based on tensor. Chin J Aeronaut 2005;18(4):326–35.
13. Wang YG, Qu XJ. Modeling decision-making aiding system for carrier launching at proper times. Acta Aeronaut Astronaut Sin 2009;30(11):2066–71 in Chinese.

CITATION

Wang Yangang, Wang Weijun, Qu Xiangju, Multi-body dynamic system simulation of carrier-based aircraft ski-jump takeoff, Chinese Journal of Aeronautics, Volume 26, Issue 1, February 2013, Pages 104-111, ISSN 1000-9361, http://dx.doi.org/10.1016/j.cja.2012.12.007.

CHAPTER 8

Mechanism of Unconventional Aerodynamic Characteristics of An Elliptic Airfoil

Wei Sun , Zhenghong Gao, , Yiming Du, Fang Xu

School of Aeronautics, Northwestern Polytechnical University, Xi'an 710072, China

ABSTRACT

The aerodynamic characteristics of elliptic airfoil are quite different from the case of conventional airfoil for Reynolds number varying from about 10^4 to 10^6. In order to reveal the fundamental mechanism, the unsteady flow around a stationary two-dimensional elliptic airfoil with 16% relative thickness has been simulated using unsteady Reynolds-averaged Navier–Stokes equations and the $\gamma - Re_{\theta t}$ transition turbulence model at different angles of attack for flow Reynolds number of 5×10^5. The aerodynamic coefficients and the pressure distribution obtained by computation are in good agreement with experimental data, which indicates that the numerical method works well. Through this study, the mechanism of the unconventional aerodynamic characteristics of airfoil is analyzed and discussed based on the computational predictions coupled with the wind tunnel results. It is considered that the boundary layer transition at the leading edge and the unsteady flow separation vortices at the trailing edge are the causes of the case. Furthermore, a valuable insight into the physics of how the flow behavior affects the elliptic airfoil's aerodynamics is provided.

INTRODUCTION

As the elliptic airfoil is applied on canard rotor/wing (CRW) aircraft, [1] and [2] more and more attention has been paid to the performance of this kind of airfoil in relatively low-Reynolds-number flows in recent years. In practice, people are especially interested in the elliptic airfoil with relatively large thickness. Kwon and Park[3] conducted an experimental study of flow over an elliptic airfoil with 16% relative thickness for the Reynolds number of 3×10^5 and found that its aerodynamic characteristics were very different from the case of conventional airfoil. Furthermore, in order to examine the influence of the Reynolds numbers, Zhan et al.[4] performed a series of experimental studies of flow over the same elliptic airfoil for a range of Reynolds number from 5×10^5 to 2.5×10^6 in the low speed wind tunnel in Northwestern Polytechnical University, by varying the wind speed from 10 m/s to 50 m/s. They also found the unconventional aerodynamic characteristics of elliptic airfoil at the Reynolds number of 5×10^5. Firstly, lift coefficient C_L increased nonlinearly with the angle of attack α, while at small angles of attack, the lift increased fast as α increased; secondly, unlike the conventional symmetrical airfoil, the minimum drag coefficient C_D was obtained at $\alpha = 4°$ rather than $\alpha = 0°$; thirdly, the variation of the pitching moment about the quarter chord was very irregular and severe and two inflection points were found in the pitching moment coefficient C_m curve. The numerical results obtained from Reynolds-averaged Navier–Stokes (RANS) equations coupled with S–A [5] and κ–ω shear stress transport (SST) [6] fully turbulence models show apparent discrepancies compared with experimental data, which can be seen clearly in Fig. 1. It indicates that these unconventional aerodynamic characteristics are difficult to capture using traditional method.

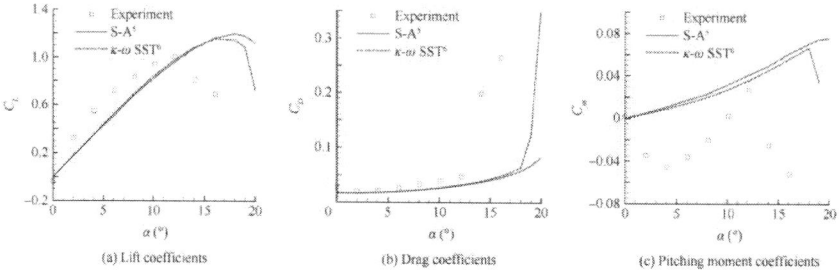

(a) Lift coefficients   (b) Drag coefficients   (c) Pitching moment coefficients

Figure 1. Predicted lift, drag and pitching moment coefficients using fully-turbulent computations against experimental data.

The flow separation commonly occurs in engineering practices.[7] In aviation, the designers always try to avoid separation or control it on aircraft surface.[8] The flow past an elliptic airfoil has been studied as a typical example of flows around blunt body since a long time ago because of its significance in fundamental flow physics. Many studies have been accomplished, some of which are experimental,[9 and 10] while the majority

of which are numerical and mostly at low Reynolds numbers.
11, 12 and 13 Unlike general airfoils, the typical characteristic of an elliptic airfoil is the blunt trailing edge, which can cause flow separation and vortex shedding to form Karman vortex street aft of the airfoil. At small angles of attack, the boundary layer is primarily laminar over the airfoil surface, but as α increases, the laminar separation bubble 14, 15 and 16 may form near the leading edge on the suction surface of the airfoil, which will result in laminar–turbulent transition. Fig. 2 shows a schematic diagram of the typical flow field structure of an elliptic airfoil. The flow separation near the blunt trailing edge and the transition inside the boundary layer have a great influence on the flow field and aerodynamic characteristics of elliptic airfoil and also pose huge challenges for computational fluid dynamics (CFD) simulation.

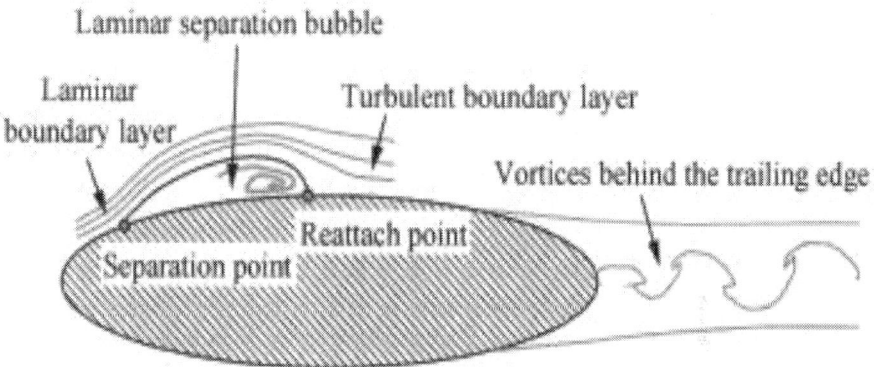

Figure 2. Schematic of flow field of elliptic airfoil.

In order to reveal the mechanism of unconventional aerodynamic characteristics exhibited by the elliptic airfoil, a numerical simulation method is established by solving the two-dimensional compressible unsteady Reynolds-averaged Navier–Stokes (URANS) equations. A four-equation transition-sensitive turbulence model is used to close the governing equations. Numerical simulations have been performed on an elliptic airfoil with 16% relative thickness for the Reynolds number of 5×10^5. Combining with the experimental data from wind tunnels, the unconventional aerodynamic characteristics are investigated.

COMPUTATION SCHEME

Governing equation

In order to simulate the unsteady vortices in the flow field, the two-dimensional compressible URANS equations are chosen as the governing equations. The non-dimensional form of the equations in Cartesian coordinates can be written as follows:

$$\frac{\partial Q}{\partial t} + \frac{\partial (E - E_v)}{\partial x} + \frac{\partial (F - F_v)}{\partial y} = 0 \tag{1}$$

where $Q=[\rho,\rho u,\rho v,e]^T$ denotes the conservative variables, ρ,u,v,e denote density, components of velocity vector and total energy per unit volume respectively, E,F denote the convective flux while E_v,F_v denote the viscous flux, whose detailed expressions are

$$\begin{cases} E = \left[\rho u,\ \rho u^2 + p,\ \rho uv,\ (e + p)u\right]^T \\ F = \left[\rho v,\ \rho uv,\ \rho v^2 + p,\ (e + p)v\right]^T \\ E_v = \left[0,\ \tau_{xx},\ \tau_{xy},\ \theta_x\right]^T \\ F_v = \left[0,\ \tau_{yx},\ \tau_{yy},\ \theta_y\right]^T \end{cases} \tag{2}$$

The viscous shear stress τ and the heat fluxes θ are in the form of

$$\begin{cases} \tau_{xx} = 2\mu u_x - \frac{2}{3}\mu(u_x + v_y) \\ \tau_{xx} = 2\mu v_y - \frac{2}{3}\mu(u_x + v_y) \\ \tau_{xy} = \tau_{yx} = \mu(u_x + v_y) \\ \theta_x = u\tau_{xx} + v\tau_{xy} + \kappa T_x \\ \theta_y = u\tau_{yx} + v\tau_{yy} + \kappa T_y \end{cases} \tag{3}$$

where κ is the coefficient of thermal conductivity and the total viscosity μ is calculated as $\mu=\mu_l+\mu_t$, where μ_l is molecular viscosity calculated by Sutherland law and μ_t is eddy viscosity determined by turbulence model.

The $\gamma - \overline{Re}_{\theta t}$ transition model[17, 18] and [19] based only on local variables is adopted to simulate laminar–turbulent transition inside the boundary layer. The $\gamma - \overline{Re}_{\theta t}$ transition model was previously developed based on the $\kappa-\omega$

SST turbulence model to resolve the laminar–turbulent transition by solving two additional transport equations for the turbulent intermittency γ and the transition onset momentum-thickness Reynolds number $Re_{\theta t}$ which read in the matrix form as follows:

$$\frac{\partial}{\partial t}\begin{bmatrix} \rho\gamma \\ \rho Re_{\theta t} \end{bmatrix} + \frac{\partial}{\partial x_j}\begin{bmatrix} \rho u_j\gamma \\ \rho u_j Re_{\theta t} \end{bmatrix} - \frac{\partial}{\partial x_j}\begin{bmatrix} \left(\mu_1 + \frac{\mu_t}{\sigma_f}\right)\frac{\partial \gamma}{\partial x_j} \\ \sigma_{\theta t}(\mu_1 + \mu_t)\frac{\partial Re_{\theta t}}{\partial x_j} \end{bmatrix} = \begin{bmatrix} P_\gamma - E_\gamma \\ P_{\theta t} \end{bmatrix}$$

(4)

The constants for the equations are σ_f=1.0 and $\sigma_{\theta t}$=2.0.

There is an effective turbulent intermittency γ_{eff} to adjust the original production and destruction terms P_κ and D_κ in the κ equation for the κ–ω SST model, which performs as follows:

$$\begin{cases} \overline{P}_k = \gamma_{eff}P_k \\ \overline{D}_k = \min\left(\max(\gamma_{eff}, 0.1), 1.0\right)D_k \\ \gamma_{eff} = \max(\gamma, \gamma_{sep}) \\ \gamma_{sep} = \min\left(s_1 \max\left[0, \left(\frac{Re_v}{3.235Re_{\theta c}}\right) - 1\right]F_{reattach}, 2\right)F_{\theta t} \\ F_{reattach} = \exp\left(-(R_T/20)^4\right) \\ s_1 = 2.0 \end{cases}$$

(5)

Here, the term γ is local turbulent intermittency and γ_{sep} the turbulent intermittency for considering the separation induced transition; Re_v is the vorticity Reynolds number and $Re_{\theta c}$ the critical momentum thickness Reynolds number; R_T is the viscous ratio and $F_{\theta t}$ the blending function from the $Re_{\theta t}$ equation.

The flow solver used in this paper is an in-house multi-block RANS solver. In this code, temporal marching method is implicit lower–upper symmetric-Gauss–Seidel (LU-SGS)[20] with sub-iterations in pseudo time, which has 2nd-order precision in unsteady calculations. The spatial discretization scheme for convection terms is Roe[21] scheme with Harten's entropy correction. 3rd-order monotonic upstream-centered scheme for conservation laws (MUSCL) reconstruction method is used to increase the Roe scheme to 2nd-order. The viscous fluxes are discretized with 2nd-order centered schemes. In order to overcome the difficulty in solving compressible equations for low Ma flows, Weiss and Smith preconditioning matrix [22] is introduced.

For turbulence model equations, implicit LU-SGS method is also operated as time marching scheme, which is similar to the discretization of governing equations. Convection terms therein are discretized by 1st-order upwind schemes, while production terms and destruction terms are treated explicitly and implicitly respectively to increase the numerical robustness.

Computational grid

According to the characteristics of the flow field, the computational domain is partitioned into three subzones: the boundary layer region, the wake region and the outer region. To ensure that the boundary location does not influence the flow, the far field boundaries for upstream, the top and bottom boundaries are placed at a distance of 10 chord lengths c from the elliptic airfoil, while the boundaries for downstream are at a distance of 20 chord lengths. In order to accurately simulate the flow inside the boundary layer, a structured O-type grid is generated in the boundary layer region: a total of 201 grid points are distributed on the airfoil surface and clustered near the leading and trailing edges. The first point in the viscous layer is 1×10^{-5} chord unit away from the wall to ensure that the y^+ value is less than unity over the entire airfoil surface. The mesh in boundary layer increases uniformly from the wall with a growth rate of 1.1. In order to resolve vortex structures in the wake region, 150 grid points are adopted there. An orthogonal C-type grid is employed in the outer region, which finally generates 70000 cells in total for simulation. A grid independent study is conducted by refining the grid in both boundary layer and wake region. No considerable changes were observed when the grid was refined and all the computations were performed on the grid described above. The grid topology and the computational grid generated are shown in Fig. 3.

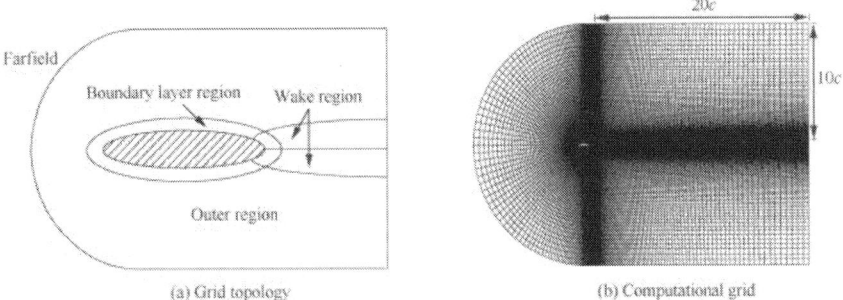

(a) Grid topology   (b) Computational grid

Figure 3. Schematic of grid topology and computational grid.

RESULTS AND ANALYSES

A non-dimensional time step size, $\Delta t = t/T$, of 0.01 is used in the simulation so as to accurately capture the unsteady flow characteristic, where $T = c/V_\infty$ is the characteristic time scale based on the ratio of chord length to the free stream velocity V_∞. An independent study of time step size is conducted with time steps smaller than 0.01, and no considerable changes are observed in results. The uniform initial condition is applied to all the unsteady simulations. The unsteady calculations are set to a total of 5000 time steps with 20 sub-iterations per time step, and it is obvious that 5000 time steps is enough to eliminate the influence of initial value and capture the features of the unsteady flow.

Since the experimental data given by the wind tunnel test in Ref.[2] are time-averaged, in order to validate the numerical method and conduct a further analysis in the present study, the time-averaged solution is obtained by taking the average over the last 3 oscillation cycles. In Fig. 4, the time-averaged lift, drag and moment coefficients are plotted in comparison with the experimental data. In order to get a clearer image of the differences between the elliptic airfoil and conventional airfoil with sharp trailing edge, NACA0016 airfoil under the same condition is chosen for comparison. It is clearly seen that the lift curve slope of the NACA0016 airfoil remains constant in linear range, the minimum drag coefficient is obtained at $\alpha = 0°$, and the pitching moment coefficient about the quarter chord is nearly constant before the airfoil stall occurs. As previously noted, the aerodynamic characteristics of elliptic airfoil are quite different from those of conventional one.

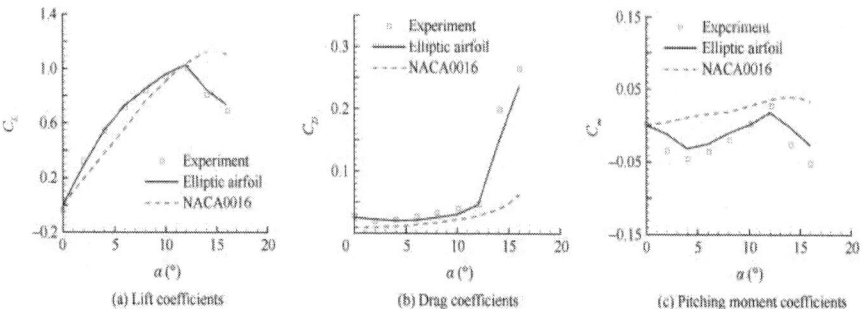

Figure 4. Experimental data for elliptic airfoil and comparison of aerodynamic force and moment coefficients between elliptic airfoil and NACA0016 airfoil.

In general, the numerical results are in good agreement with the experimental data. The numerical method is proved to be reliable since the unconventional aerodynamic characteristics of elliptic airfoil, such as the high lift slope at small angles of attack, the nonlinear segment along C_L distribution, the large C_D value at $\alpha = 0°$ and the irregular variation along C_m distribution are accurately predicted. However, there is one point that needs attention: the computation result of C_L at $\alpha = 0°$ is zero while the experimental result is not zero, which implies that an important aspect of the experimental geometry has not been adequately modeled in the computational technique or that there is error in the experimental value; obviously, the later one is reasonable.

In order to gain a deep insight into the unconventional aerodynamic characteristics of the elliptic airfoil, the analysis is presented in the following three different aspects.

Characteristics of lift

The aerodynamic performance of an airfoil can be studied easier by referring to the pressure distribution over itself. In order to find out the reason for the high lift and high lift slope at small angles of attack and why a shift exists in the lift slope between $\alpha = 4°$ and $\alpha = 6°$, the time-averaged pressure coefficient C_p distribution is computed for $\alpha = 2°$, $4°$, $6°$, $8°$. The results are in good agreement with the experimental data (shown in Fig. 5), which further validates the numerical method.

It is very interesting to find out that there is a sudden increase in C_p near the trailing edge for $\alpha < 6°$. This local pressure recovery area makes the adverse pressure gradient mild over the most part of the upper surface. The C_p curve over the upper surface is flat, as a result the elliptic airfoil carries more lift than the conventional ones. In addition, the negative pressure peak rises quickly as α increases, which further enhances the lift, and also brings a high lift slope. When α is up to $6°$, the local pressure recovery area appears in the place near the leading edge after the negative pressure peak, forming a slight pressure plateau region. This characteristic pressure distribution is indicative of the formation of laminar separation bubble. The sudden increase in C_p after the bubble leads to a relatively heavy loss in lift. And therefore the lift growth rate decreases.

Actually, to figure out what must be responsible for these unconventional lift characteristics, we need to figure out what happened in the flow around the elliptic airfoil. The time-averaged skin friction coefficient C_{fx} distribution and turbulent kinetic energy (TKE) distribution around the airfoil for different angles of attack are shown in Fig. 6. It can be observed that at $\alpha < 6°$, the C_{fx} distribution curve declines smoothly and TKE is zero except in the separated flow region near the blunt trailing edge, which means the boundary layer is laminar over most of the airfoil. At $\alpha = 6°$, skin friction coefficient C_{fx} becomes negative along chord wise firstly, indicating that the laminar boundary layer separates, and then C_{fx} increases rapidly to be positive, indicating that the laminar–turbulent transition occurs and the turbulent boundary layer reattaches to the airfoil surface. Higher TKE can be observed in the boundary layer beyond the transition point, and the reattached turbulent boundary layer is much more energetic. In summary, the wide range of laminar flow over the elliptic airfoil is the real root cause of the high lift and high lift slope at small angles of attack. Once the boundary layer transition happens, the unconventional lift characteristic disappears.

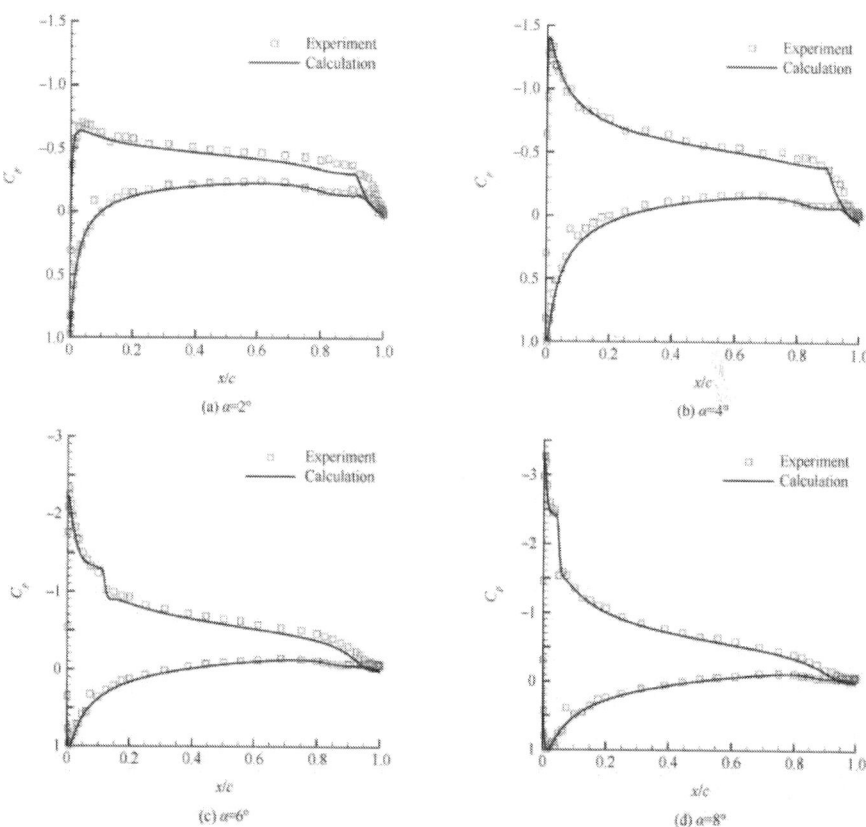

Figure 5. Time-averaged surface pressure coefficients distribution in comparison with experimental results.

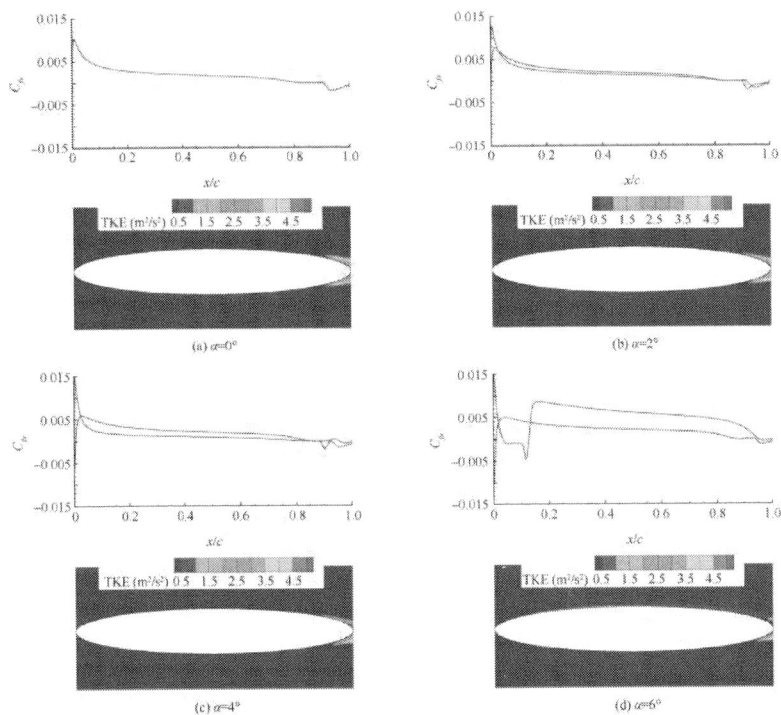

Figure 6. Time-averaged skin friction coefficients distribution and turbulent kinetic energy distribution.

Characteristics of drag

At $\alpha < 6°$, we find that the drag coefficient decreases as the flow angle increases and the minimum value of C_D is obtained at $\alpha = 4°$. It is very unconventional because the research object in this study is a symmetrical airfoil. Fig. 7 shows the components of the drag, pressure drag C_{Dp} and frictional drag C_{Df}, as functions of angles of attack. Since the pressure drag is the main component, the characteristics of drag have a close relationship with the vortices at the trailing edge. The time-averaged vortex structure with mean stream line at small angles of attack is illustrated in Fig. 8. As the angle of attack increases, the streamlines packed in the recirculation region will be denser, and therefore the cross-sectional areas will be smaller which results in smaller pressure drag. However, after the laminar–turbulent transition occurs inside the boundary layer at $\alpha = 6°$, the frictional drag increases rapidly due to the turbulent boundary layer on the upper surface. Hence, C_D begins to increase beyond the angle of attack for minimum drag, which is around $4°$.

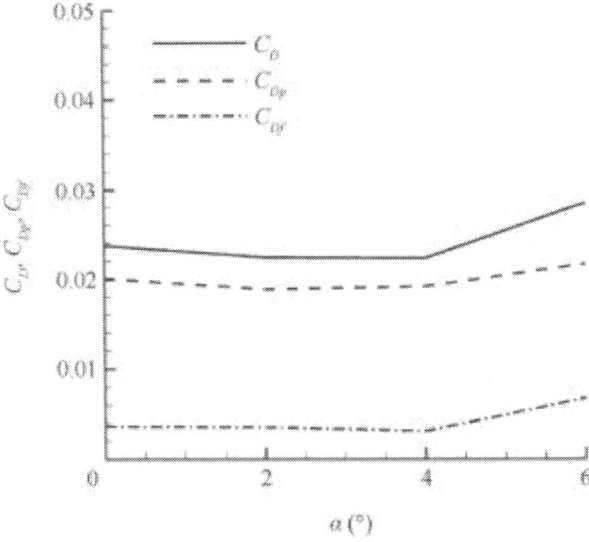

Figure 7. Components of drag, pressure drag and frictional drag coefficients at small angles of attack.

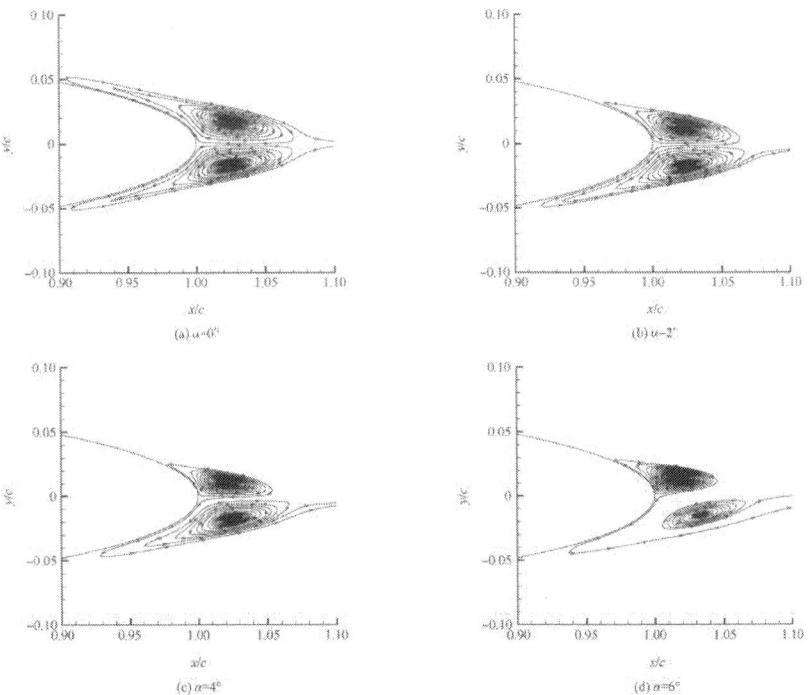

Figure 8. Time-averaged vortex structure with stream line at trailing edge.

Characteristics of moment

It is known that in the case of conventional airfoil, the pitching moment coefficient C_m about the quarter chord is nearly constant in the linear lift range. By contrast, the pitching moment coefficient of the elliptic airfoil varies very unconventionally with the angle of attack, which is shown in Fig. 4. At small angles of attack, C_m is negative, and this nose-down moment becomes more extreme as the angle of attack increases. After $\alpha = 6°$, this tendency shows a sign of abating and the first inflection point shows up. The pitching moment begins to increase beyond that angle of attack. When α is up to $10°$, C_m becomes positive, which means it changes from nose-down moment to nose-up moment. But when α exceeds $12°$, the second inflection point shows up and the pitching moment becomes nose-down moment soon after.

In order to better understand the moment characteristics, Table 1 and Table 2 give the components of moment at small angles of attack and large attack angles respectively, where C_{mp} refers to the contribution of pressure while C_{mf} refers to the contribution of viscous force on moment. Apparently, the C_{mf} is very small. At small angles of attack where $\alpha < 6°$, as shown in Fig. 5, the elliptic airfoil is a bit aft-loaded so the aft section of the airfoil produces more lift and nose-down pitching moment. At $\alpha = 6°$, laminar–turbulent transition takes place inside the boundary layer near the leading edge; although C_{mf} becomes much larger than before, it is still fairly small. The real reason that the moment begins to increase, is that the aft-loaded phenomenon disappears because of transition near the leading edge. While at large angles of attack where $\alpha > 12°$, the C_{mf} becomes negative, which must be caused by reverse flow over the upper surface. This can be confirmed by Fig. 9, velocity contour of the elliptic airfoil at $\alpha = 14°$. The separated boundary layer fails to reattach to the suction surface and massive flow separation can be observed. It is this massive flow separation that leads to the undesirable change of moment.

Table 1. Components of moment coefficients at small angles of attack.

α (°)	C_{mp}	C_{mf}	C_m
2	−0.0222	0.000048	−0.0221
4	−0.0377	0.000085	−0.0376
6	−0.0227	0.00046	−0.0223

Table 2. Components of moment coefficients at large angles of attack.

α (°)	C_{mp}	C_{mf}	C_m
10	−0.0007	0.00048	−0.0003
12	0.0145	0.00041	0.0149
14	−0.0019	−0.00020	−0.0021

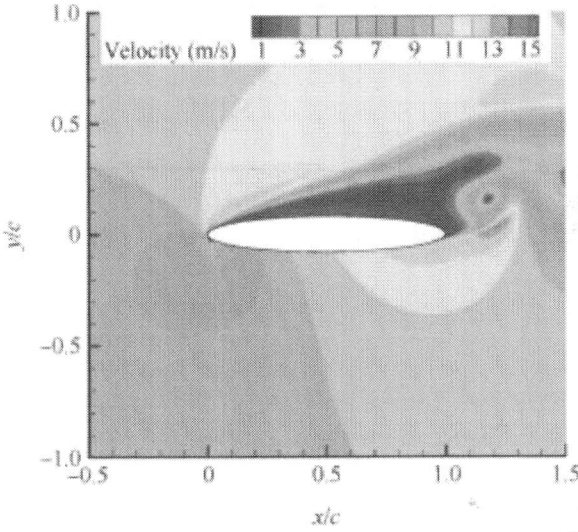

Figure 9. Instantaneous velocity contour at $\alpha = 14°$.

CONCLUSIONS

(1) For the flows at relatively low Reynolds number, laminar–turbulent boundary layer transition and separation flow behind the trailing edge play a predominant role in determining aerodynamic characteristics of the elliptic airfoil. Since the flow is essentially unsteady, the URANS method and the transition turbulent model are indeed necessary to accurately simulate the flow field.

(2) The wide range of laminar flow over the elliptic airfoil is the root cause of the high lift slope at small angles of attack. Once the boundary layer transition takes place near the leading edge, the local pressure recovery area leads to great loss to lift and high lift slope characteristic vanishes.

(3) As α increases from $0°$ to $4°$, the main component of drag is pressure drag, the mean cross-sectional area of the separation vortices decreases which leads to the decreases in pressure drag, the frictional drag shows little change such that the total drag decreases. When laminar–turbulent transition occurs at $\alpha = 6°$, the frictional drag increases dramatically, and the pressure drag increases slightly, so that the total drag increases.

(4) At small angles of attack where boundary layer is laminar over the elliptic airfoil, the aft-load carries nose-down pitching moment; at $\alpha = 6°$ where the boundary layer transition occurs, the loss of lift at the aft part of airfoil results in increased pitching moment; while at large angles of attack where the airfoil stalls, the massive flow separation over the suction surface of airfoil is responsible for irregular variation of pitching moment.

ACKNOWLEDGEMENTS

The authors thank the anonymous reviewers for their critical and constructive review of the manuscript. The current work was financially supported by National Natural Science Foundation of China (No. 11372254).

REFERENCES

1. Mitchell CA, Vogel BJ. The canard rotor wing (CRW) aircraft–a new way to fly. 2003. Report No.: AIAA-2003-2517.
2. Deng YP, Gao ZH, Zhan H. Development and key technologies of the CRW. Flight Dyn 2006;24(3):1–4 [Chinese].
3. Kwon K, Park SO. Aerodynamic characteristics of an elliptic airfoil at low Reynolds number. J Aircraft 2005;42(6):1642–4.
4. Zhan H, Deng YP, Gao ZH. Investigation on aerodynamics performance of elliptic airfoil at low speed. Aeronaut Comput Tech 2008;38(3):25–7 [Chinese].
5. Spalart PR, Allmaras SR. A one–equation turbulence model for aerodynamic flows. 1992. Report No.: AIAA-1992-0439.
6. Menter FR. Two-equation eddy viscosity turbulence models for engineering applications. AIAA J 1994;32(8):1598–605.

7. Xu F, Gao ZH, Ming X, Xia L, Wang YH, Sun W, et al. The optimization of the backward-facing step flow control with synthetic jet based on experiment. Exp Therm Fluid Sci 2015;64:94–107.

8. Zhao GQ, Zhao QJ. Parametric analyses for synthetic jet control on separation and stall over rotor airfoil. Chin J Aeronaut 2014;27(5):1051–61.

9. Zahm AF, Smith RH, Louden FA. Forces on elliptic cylinders in uniform air stream. Washington, D. C.: National Advisory Committee for Aeronautics; 1929. Report No.: 289.

10. Schubauer GB. Air flow in boundary layer of an elliptic cylinder. Washington, D.C.: National Advisory Committee for Aeronautics; 1938. Report No.: 652.

11. Kim MS. Unsteady viscous flow over elliptic cylinders at various thickness with different Reynolds numbers. Reston: AIAA; 2005. Report No.: AIAA-2005-5130.

12. Nair MT, Sengupta TK. Unsteady flow past elliptic cylinders. J Fluid Struct 1997;11(6):555–95.

13. Chitta V, Walters DK. Prediction of aerodynamic characteristics of an elliptic airfoil at low Reynolds number. Proceedings of the ASME 2012 fluids engineering division summer meeting. Rio Grande: ASME; 2012. p. 1297–380.

14. Diwan SS, Ramesh ON. Laminar separation bubbles: dynamics and control. Sadhana 2007;32:103–9.

15. Jahanmiri M. Laminar separation bubble: its structure, dynamics and control. Goteborg: Chalmers University of Technology; 2011. Report No.: 201106.

16. Marxen O, Lang M, Rist U. Vortex formation and vortex breakup in a laminar separation bubble. J Fluid Mech 2013;728:58–9.

17. Langtry RB, Menter FR. Correlation-based transition modeling for unstructured parallelized computational fluid dynamics codes. AIAA J 2009;47(12):2894–906.

18. Langtry RB. A correlation-based transition model using local variables for unstructured parallelized CFD codes [dissertation]. Stuttgart: University Stuttgart; 2006.

19. Zhang XD, Gao ZH. A numerical research on a compressibilitycorrelated Langtry's transition model for double wedge boundary layer flows. Chin J Aeronaut 2011;24(3):249–57.

20. Yoon S, Jameson A. Lower-upper symmetric-Gauss–Seidel method for the Euler and Navier–Stokes equations. AIAA J 1988;26(9):1025–6.

21. Roe PL. Approximate Riemann solvers, parameter vectors, and difference schemes. J Comput Phys 1981;43(2):357–72.

22. Weiss JM, Smith WA. Preconditioning applied to variable and constant density flow. AIAA J 1995;33(11):2050–7.

CITATION

Wei Sun, Zhenghong Gao, Yiming Du, Fang Xu, Mechanism of unconventional aerodynamic characteristics of an elliptic airfoil, Chinese Journal of Aeronautics, Volume 28, Issue 3, June 2015, Pages 687-694, ISSN 1000-9361, http://dx.doi.org/10.1016/j.cja.2015.03.009.

CHAPTER 9

Unsteady Aerodynamic Modeling at High Angles of Attack Using Support Vector Machines

Qing Wang[1], Weiqi Qian[2], Kaifeng He[1,2]

[1]State Key Laboratory of Aerodynamics, Mianyang 621000, China

[2]Computational Aerodynamics Institute, China Aerodynamics Research and Development Center, Mianyang 621000, China

ABSTRACT

Accurate aerodynamic models are the basis of flight simulation and control law design. Mathematically modeling unsteady aerodynamics at high angles of attack bears great difficulties in model structure determination and parameter estimation due to little understanding of the flow mechanism. Support vector machines (SVMs) based on statistical learning theory provide a novel tool for nonlinear system modeling. The work presented here examines the feasibility of applying SVMs to high angle-of-attack unsteady aerodynamic modeling field. Mainly, after a review of SVMs, several issues associated with unsteady aerodynamic modeling by use of SVMs are discussed in detail, such as selection of input variables, selection of output variables and determination of SVM parameters. The least squares SVM (LS-SVM) models are set up from certain dynamic wind tunnel test data of a delta wing and an aircraft configuration, and then used to predict the aerodynamic responses in other tests. The predictions are in good agreement with the test data, which indicates the satisfying learning and generalization performance of LS-SVMs.

INTRODUCTION

Unsteady aerodynamics at high angles of attack plays an increasingly important part in modern aircraft design. In high angle-of-attack maneuvers, the flow field around the aircraft is extremely complex and the aerodynamics shows strong nonlinearity and unsteadiness. As a result, the conventional aerodynamic database, composed of static test data, dynamic derivatives and rotary-balance data, does not meet the requirements of flight simulation and control law design. The database needs to involve dynamic test data. Unfortunately, the aerodynamic characteristics at high angles of attack, especially in post-stall maneuvers, cannot be predicted simply by interpolation among limited test data, as done at pre-stall angles of attack. A feasible solution is to set up aerodynamic models, which describe the dependence of aerodynamics upon the motion history, from a certain number of static and dynamic wind tunnel test data.

Currently, the researches on high angle-of-attack unsteady aerodynamic modeling evolve in two directions: mathematic methods and artificial intelligent methods. The developed mathematic models include those in form of generalized aerodynamic derivatives,[1] nonlinear indicial response,[2, 3 and 4] internal state-space,[5 and 6] differential equations,[7, 8, 9 and 10] hybrid representation of nonlinear indicial response and internal state-space,[11] flow incidence rate,[12] etc. They are based on the understanding of physical phenomenon and mechanism. The available intelligent methods include fuzzy logic (FL)[13, 14 and 15] and neural networks (NNs),[16, 17, 18, 19 and 20] which are suitable to black-box system modeling especially. In addition, reduced order models of nonlinear and unsteady aerodynamics, based on indicial response functions, have been developed.[21, 22 and 23] These functions can be estimated via analytical, experimental or computational methods. The analytical solutions are limited only to two-dimensional airfoils in incompressible and inviscid flows.[21] Experimental tests are practically nonexistent for indicial response functions. CFD calculations have recently been used to determine indicial response functions for the given aircraft configurations.[22 and 23]

A large number of wind tunnel tests[24, 25 and 26] and CFD simulations[27, 28 and 29] have been conducted to study unsteady aerodynamics of maneuvering aircraft at high angles of attack. One has acquired some knowledge about the effects of reduced frequency, amplitude, and mean angle of attack on unsteady aerodynamics in forced-oscillations. However, many problems in the area of unsteady flow mechanism at high angles of attack, particularly in the case of multiple degree-of-freedom (DOF) coupling motions, have not been solved yet, which leads to great difficulties in model structure determination and parameter estimation when mathematically modeling unsteady aerodynamics. In this case, intelligent methods are gaining popularity. They let computers learn the available static and dynamic wind tunnel test data and then predict the aerodynamic responses of aircraft in flight. In intelligent methods, how to determine the optimal model structure is a critical problem, which has not been solved perfectly by FL and NNs. Support vector machines (SVMs), a new type of statistical learning strategy, embody the structural risk minimization (SRM) principle, which has been shown to be superior to the traditional empirical risk minimization (ERM) principle, employed by conventional FL and NNs.[30, 31 and 32] Therefore, SVMs exhibit more excellent empirical performance than FL and NNs. Another attractive feature of SVMs is the global optimality. By introducing a nonlinear map from input space to feature space, SVMs transform a nonlinear system modeling problem to a quadratic programming, which can achieve global minimum.

This paper presents a pioneer study of using SVMs to model high angle-of-attack unsteady aerodynamics. After a review of SVMs and least squares SVMs (LS-SVMs), an extension of the standard SVMs, a simulation experiment of a two-dimensional nonlinear system is performed to validate the empirical performance of SVMs. Several issues associated with application of SVM method in unsteady aerodynamic modeling field are discussed in detail, such as selection of input variables, selection of output variables, and determination of SVM parameters. The LS-SVM method is applied to aerodynamic modeling of a pitching delta wing and a rolling aircraft configuration. The satisfying learning and generalization performance exhibited by the applications indicates the feasibility of applying SVMs to high angle-of-attack unsteady aerodynamic modeling.

SVMs FOR NONLINEAR SYSTEM MODELING

SVMs, a novel tool in the area of machine learning, were first proposed by Vapnik[33] in 1995. Originally, they were developed for pattern recognition problems. Recently, they have been successfully extended to nonlinear function approximation and nonlinear system modeling.

SVM regression

The SVMs used in system modeling are called SVM regression, or support vector regression (SVR) in short. For a problem of multi-input single-output (MISO) nonlinear system modeling, SVMs approximate the nonlinear function by a linear regression:

$$y = f(x) = w^{\mathrm{T}}\varphi(x) + b \quad x \in \mathbf{R}^m, y \in \mathbf{R} \tag{1}$$

in a high-dimensional feature space F. Here $\varphi(x)$ denotes a nonlinear transformation from input space $\mathbf{R}^m$ to feature space F, w is weighting vector, and b is bias.

Suppose that a finite number set of sample data $\{(x_i, y_i), i = 1, 2, \ldots, n\}$ have been obtained by experimental measurement. If all the training data can be fitted by the function Eq. (1) with ε precision, then

$$\begin{cases} y_i - w^T \varphi(x_i) - b \leqslant \varepsilon \\ w^T \varphi(x_i) + b - y_i \leqslant \varepsilon \end{cases} \quad i = 1, 2, \cdots, n \tag{2}$$

Sometimes, however, this may not be the case, or we may also want to allow for some error. One can introduce slack variables ξ and ξ^* to cope with otherwise infeasible constrains:

$$\begin{cases} y_i - w^T \varphi(x_i) - b \leqslant \varepsilon + \xi_i \\ w^T \varphi(x_i) + b - y_i \leqslant \varepsilon + \xi_i^* \end{cases} \quad i = 1, 2, \cdots, n \tag{3}$$

The SRM principle yields the optimization goal:

$$\min_{w, b, \xi, \xi^*} J = \frac{1}{2} \|w\|^2 + c \sum_{i=1}^{n} (\xi_i + \xi_i^*) \tag{4}$$

where c is penalty factor and a pre-specified constant determining the training error and the regression function flatness.

Using the Lagrange function method together with the dual variables to find the solution of the above problem can lead to a quadratic programming (QP) problem:

$$\begin{cases} \max_{a, a^*} J = -\frac{1}{2} \sum_{i,j=1}^{n} (a_i - a_i^*)(a_j - a_j^*) K(x_i, x_j) - \varepsilon \sum_{i=1}^{n} (a_i + a_i^*) \\ \qquad\qquad + \sum_{i=1}^{n} y_i (a_i - a_i^*) \\ \text{s.t.} \begin{cases} \sum_{i=1}^{n} y_i (a_i - a_i^*) = 0 \\ a_i, a_i^* \in [0, c] \end{cases} \end{cases} \tag{5}$$

where a_i and a_i^* are the Lagrange multipliers; $K(x_i, x_j)$ is called kernel function. Its value is equal to the inner product of two vectors x_i and x_j in the feature space $\varphi(x_i)$ and $\varphi(x_j)$, i.e., $K(x_i, x_j) = \varphi(x_i) \cdot \varphi(x_j)$.

Solving the above QP problem with inequality constrains, the Lagrange multipliers a_i and a_i^* can be determined. Then, the weighting vector w and the bias b can be derived from Karush–Kuhn–Tucker's (KKT's) condition.

$$
\begin{cases}
w = \sum_{i=1}^{n}(a_i - a_i^*)\varphi(x_i) \\
b = \frac{1}{n}\sum_{j=1}^{n}\left[y_i - \sum_{i=1}^{n}(a_i - a_i^*)K(x_i, x_j) + \varepsilon \ \mathrm{sgn}(a_j - a_j^*) \right]
\end{cases}
\tag{6}
$$

Therefore, the line regression Eq. (1) becomes the following explicit form.

$$
y = f(x) = \sum_{i=1}^{n}(a_i - a_i^*)K(x, x_i) + b
\tag{7}
$$

Based on the nature of the corresponding QP, in general, only a number of coefficients $a_i - a_i^*$ will be assumed as non-zero, and the data points associated with the pair can be referred to as support vectors (SVs).

Fig. 1 shows the topologic structure of SVMs. In Fig. 1, $x = [x_1, x_2, \ldots, x_m]^T$.

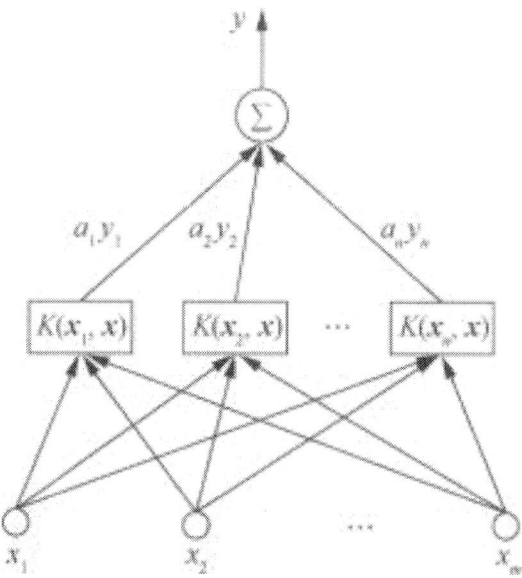

Figure 1. Topologic structure of SVMs.

LS-SVMs

The goal function of SVMs is convex and thus has only one extreme value. However, dimension disaster will arise if the number of training samples is very large, which may result in the optimization algorithm too complex to be carried out. As an extension of the standard SVMs, LS-SVMs are proposed by Suykens and Vandewalle[34] in 1999. The algorithm complexity of LS-SVMs is reduced down greatly by solving linear algebraic equations instead of QP. They have been extensively applied to function approximation and system modeling.

In LS-SVMs, the linear term in the goal function Eq. (4) is replaced by the square term of ξ_i and the inequality constrains Eq. (3) are replaced by equality constrains. Thus, the optimization problem can be written as:

$$\begin{cases} \min_{w,b,\xi,\xi^*} J = \frac{1}{2}\|w\|^2 + \frac{1}{2}c\sum_{i=1}^{n}\xi_i^2 & i = 1,2,\cdots,n \\ \text{s.t.} \quad y_i = w^T\varphi(x_i) + b + \xi_i \end{cases} \tag{8}$$

The Lagrange function is introduced to solve the above equality-constrained optimization problem:

$$L = \frac{1}{2}\|w\|^2 + \frac{1}{2}c\sum_{i=1}^{n}\xi_i^2 - \sum_{i=1}^{n}a_i(w^T\varphi(x_i) + b + \xi_i - y_i) \tag{9}$$

From KKT's condition, one gets the equations:

$$\begin{cases} w = \sum_{i=1}^{n}a_i\varphi(x_i) \\ \sum_{i=1}^{n}a_i = 0 \\ a_i = c\xi_i \qquad i = 1,2,\cdots,n \\ w^T\varphi(x_i) + b + \xi_i - y_i = 0 \qquad i = 1,2,\cdots,n \end{cases} \tag{10}$$

After eliminating w and ξ_i, the following linear system is obtained.

$$\begin{bmatrix} 0 & 1_n^T \\ 1_n & \Omega + c^{-1}I_n \end{bmatrix}\begin{bmatrix} b \\ a \end{bmatrix} = \begin{bmatrix} 0 \\ y \end{bmatrix} \tag{11}$$

where

$$\begin{cases} \boldsymbol{y} = [y_1, y_2, \cdots, y_n]^{\mathrm{T}} \\ \boldsymbol{1}_n = [1, 1, \cdots, 1]^{\mathrm{T}} \\ \boldsymbol{a} = [a_1, a_2, \cdots, a_n]^{\mathrm{T}} \\ \Omega_{i,j} = K(\boldsymbol{x}_i, \boldsymbol{x}_j) \quad i, j = 1, 2, \cdots, n \end{cases} \tag{12}$$

Eq. (11) can be solved for the parameters a_i (i = 1,2, ... ,n) and b by use of least squares method. Therefore, the resulting LS-SVM model is given as

$$y = f(\boldsymbol{x}) = \sum_{i=1}^{n} a_i K(\boldsymbol{x}, \boldsymbol{x}_i) + b \tag{13}$$

As a result of the modifications in LS-SVMs, training requires only to solve the linear equations Eq. (11) instead of the computationally hard quadratic programming problem Eq. (5) in the standard SVMs. However, this is done at expense of losing the sparseness of solution. All the training samples act as support vectors in LS-SVMs.

A simple example

In order to validate the learning and generalizing capability of LS-SVMs, a simulation experiment is given as follows.

Consider a two-input one-output nonlinear system:

$$y = \frac{\sin^2 \pi(x_1 - x_2)}{1 + x_1^2 + x_2^2} \tag{14}$$

It constitutes a spatial curved surface. Taking 200 points on $[-1, 1] \times [-1, 1]$ stochastically, we create training sample data by the following formula.

$$\begin{cases} y = \frac{\sin^2 \pi(x_1 - x_2)}{1 + x_1^2 + x_2^2} + \eta \\ \eta \sim N(0, 0.01^2) \end{cases} \tag{15}$$

Fig. 2(a) shows the exact curved surface and the training samples.

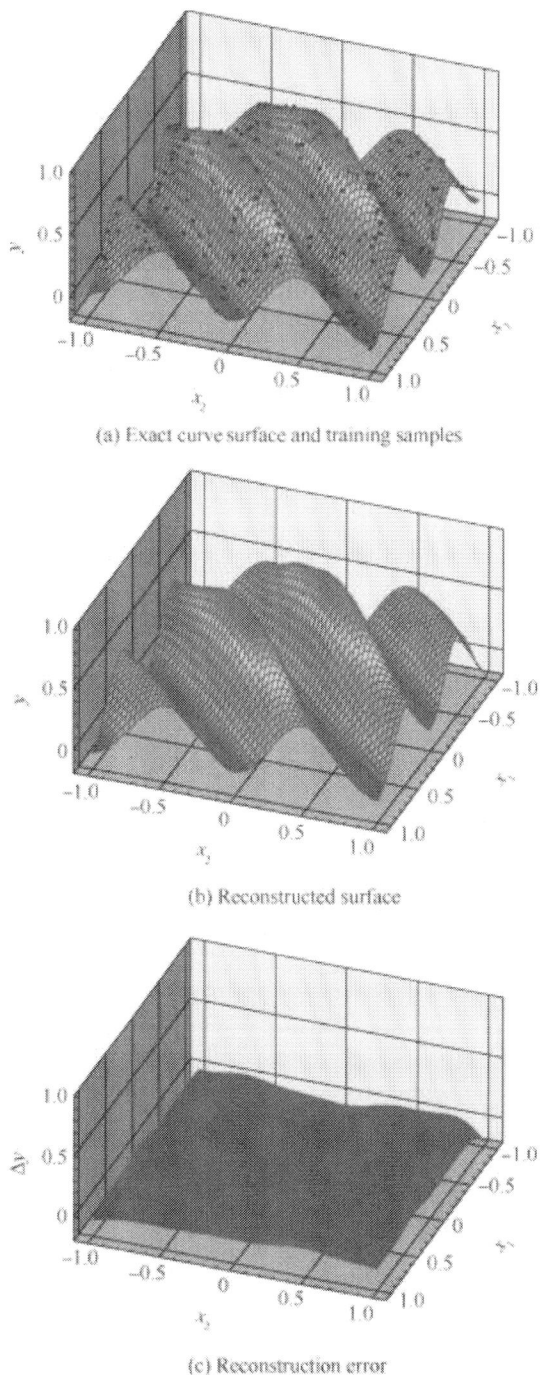

(a) Exact curve surface and training samples

(b) Reconstructed surface

(c) Reconstruction error

Figure 2. LS-SVM modeling results of a two-dimensional nonlinear system.

With the following radius base function adopted, the LS-SVM model is set up for the nonlinear system Eq. (14) from the training samples, where the penalty factor $c = 4$ and the width parameter of radius base kernel function $2\sigma^2 = 0.25$.

$$K(x, x_i) = \exp \left(-\frac{\|x - x_i\|^2}{2\sigma^2} \right)$$

(16)

where σ is kernel width. Then, the surface is reconstructed using the LS-SVM model and shown in Fig. 2(b). The reconstruction error Δy is shown in Fig. 2(c). It is seen that the reconstructed surface approximate the exact one very well. One can expect that the reconstruction error will be reduced down further with decline of the noise level in training samples and/or increase of the number of training samples.

HIGH ANGLE-OF-ATTACK UNSTEADY AERODYNAMIC MODELING METHOD BY USE OF SVMS

As mentioned previously, little understanding of unsteady flow mechanism at high angles of attack leads to great difficulties in model structure determination and parameter estimation when mathematically modeling aerodynamics. The development of SVMs provides a novel tool for unsteady aerodynamic modeling at high angles of attack. Three issues below are involved in high angle-of-attack unsteady aerodynamic modeling by use of SVMs.

Selection of input variables

One of the important characteristics of high angle-of-attack aerodynamics is its dependence not only on the instantaneous flight states but also on their time history. Hence, the selection of input variables must enable the SVMs to incorporate the impact of motion history on aerodynamics.

The parameter of reduced frequency k is employed as one of the input variables in most of the previous researches on intelligent modeling such as FL [13, 14] and 15 and NNs. [18] For example, Ref. [13] took $(\alpha, \dot{\alpha}, \ddot{\alpha}, k, \beta, \delta_e)$ as the input variables when modeling aerodynamics of pitching oscillations, while Ref. [14] took $(\alpha, \phi, \dot{\phi}, k, \beta, \psi, \dot{\psi})$ as the input variables when modeling aerodynamics of yawing and rolling oscillations, where α is angle of attack, β is sideslip angle, ϕ is roll angle, δ_e is elevator deflection angle and ψ is yawing angle. It is noted that the reduced frequency is an unsteadiness parameter adopted specifically in forced-oscillation wind tunnel tests, which does not exist in flight tests. When the aerodynamic models set up from wind tunnel test data are used to predict aerodynamic characteristics in flight tests, an equivalent reduced frequency is needed, which is obtained by fitting a segment of flight test data. For example, the equivalent reduced frequency of pitching motion is calculated by solving an optimization problem [15]:

$$\min_{\alpha_0,\alpha_s,k,\varphi} J = \sum_{i=1}^{n}\{\alpha - [\alpha_0 + \alpha_s \cos(k\tau + \varphi)]\}^2$$

$$+ \sum_{i=1}^{n}\{\bar{\dot{\alpha}} - [-\alpha_s k \sin(k\tau + \varphi)]\}^2 \qquad (17)$$

where $\bar{\dot{\alpha}}$ is nondimensional rate of angle of attack; τ is nondimensional time; n is the pre-specified number of points for fitting. The mean angle-of-attack α_0, amplitude α_s, reduced frequency k, and phase φ are the parameters to be determined.

A series of problems arise from this processing technique: (a) The nonlinearity of the goal function may lead to local optimality. (b) Singularity may occur in some cases. As a special case, k can be any value at constant angles of attack. (c) For coupled 6-DOF motions, it is not assured whether the equivalent reduced frequency should be determined in three directions of body-axis or in two directions of angle of attack and sideslip angle. (d) The number of points for fitting may have great effects on the resulting equivalent reduced frequency. (e) The resulting equivalent reduced frequency may vary acutely along with time. Fig. 3 shows the time history of the equivalent parameter during a coupled yawing-rolling ramp motion. [18] It skips frequently in the process of ramp (0–3 s) and the reverse (6.7–9.7 s).

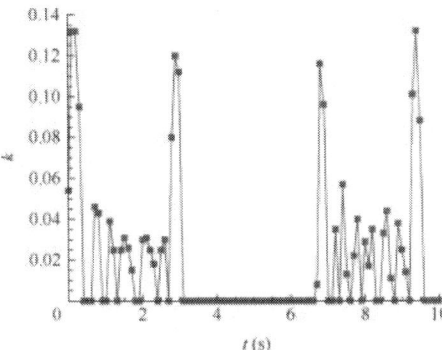

Figure 3. Equivalent reduced frequency of coupled yawing-rolling ramp motion.[18]

As a solution, several sampling points of the current and previous flight states could be employed to describe the effects of motion history. Ref.[16] took $[\alpha(\tau), \alpha(\tau-1), \alpha(\tau-2); \dot{\alpha}(\tau), \dot{\alpha}(\tau-1), \dot{\alpha}(\tau-2); ...]$ as input variables when modeling the longitudinal aerodynamics of a pitching delta wing by use of back propagation neural network (BPNN). Ref.[19] took $[\alpha(\tau), \alpha(\tau-\Delta\tau), \alpha(\tau-2\Delta\tau), \alpha(\tau-3\Delta\tau), \alpha(\tau-4\Delta\tau); \beta(\tau), \beta(\tau-\Delta\tau), \beta(\tau-2\Delta\tau), \beta(\tau-3\Delta\tau), \beta(\tau-4\Delta\tau)]$ as input variables when modeling 6-component aerodynamic coefficients by use of radius base function neural network (RBFNN). The references exhibit satisfying results.

In this way, however, two problems remain to be solved: (a) How long time before do the flight states have no effect upon the current aerodynamics? (b) How to determine the number of sampling points? As for the first question, the time length $m\cdot\Delta\tau$ can be determined according to the state-space models 5 and 6 or the differential-equation models, [7, 8, 9 and 10] to be $(1-2)\tau_1$ for instance, where τ_1 is the characteristic time constant in the mathematical models. As for the second question, the number of sampling points m can be determined appropriately according to the above time length and the mode frequencies of the aircraft, to be $[8(m\cdot\Delta\tau)\cdot\max(\bar{\omega}_{SP}, \bar{\omega}_{DR})]$ for instance, where $\bar{\omega}_{SP}$ is the nondimensional frequency of short-period mode and $\bar{\omega}_{DR}$ is that of Dutch roll mode.

Selection of output variables

In intelligent modeling of high angle-of-attack unsteady aerodynamics, the most direct output variables are aerodynamic force and moment coefficients. There is a fall in this way that the errors in dynamic test data will affect the predictions of static aerodynamics. As we know, dynamic wind tunnel test data include greater errors than static ones in general.

With a view to engineering application, aerodynamics can be decomposed as follows (pitch moment coefficient C_m as an example).

$$C_m = C_{m,st} + \Delta C_{m,dyn} \tag{18}$$

A candidate way is to model the dynamic increment $\Delta C_{m,dyn}$ instead of the aerodynamic coefficient C_m, while the static component $C_{m,st}$ takes the static wind tunnel test results. Thus, all the dynamic test data, measured in different forms of motions and even in different wind tunnels, can be put together for SVM training.

Determination of SVM parameters

The selection of kernel function and determination of SVM parameters are another important problems for aerodynamic modeling. They also have decisive effects upon the fitting precision, the generalization ability, and the training speed of SVMs.

The kernel function decides the mapping pattern from the input space to the feature space. Theoretically, the kernel can be any symmetry function satisfying Mercer's condition. The commonly used kernels include:

(1) Polynomial function: $K(x, x_i) = (x^T x_i + 1)^q$
(2) Radius base function (RBF): $K(x, x_i) = \exp[-\|x - x_i\|^2/(2\sigma^2)]$
(3) Sigmoid function: $K(x, x_i) = \tanh[r(x^T\text{-}x_i) + c]$
(4) Potential function: $K(x, x_i) = \exp[-\|x - x_i\|/(2\sigma^2)]$

There is no universal principle to guide the selection of kernel function. However, the RBF kernel is efficient for nonlinear system modeling, which is validated by a large number of simulation experiments and applications.

Having selected the kernel function (RBF kernel is adopted in unsteady aerodynamic modeling in the next section), the penalty factor c and the kernel width σ in LS-SVMs need also to be determined. The performance of LS-SVMs will be improved by adjusting the two parameters. The methods such as m-fold cross-validation and leave-one-out (LOO) can be utilized to determine these parameters. 35 [and] 36

The m-fold cross-validation is the most commonly used method to estimate the generalization error. In this method, the training sample set is first partitioned into $m$subsets with nearly the same size which do not intersect each other. For given parameters c and σ, one of the subsets is taken as the testing set at a time, the others as the training set. After the parameters a_i ($i = 1,2, \ldots ,n$) and b in the model are obtained by the training algorithm from the training set, the mean square sum of prediction errors $MSSE_i$ can be calculated for the testing set and is taken as the testing error. The average of m testing errors $MSSE = (1/m)\Sigma MSSE_i$ is taken as the generalization error. Thus, the parameters c and σ can be adjusted according to the generalization error.

For a group of parameters c and σ, totally m times of training and testing are needed to calculate the average error MSSE. Consequently, using m-fold cross-validation to optimize the parameters is computationally expensive. The higher m is, the more computation time it costs. One should select an appropriate low value of m to perform m-fold cross-validation, especially in case of large number of training samples.

For unsteady aerodynamic modeling at high angles of attack, it is important that all the data of an individual dynamic test should be taken as a unit and put into same a subset during the m-fold partitioning.

RESULTS AND DISCUSSION

Aerodynamic modeling of a pitching delta wing

The wind tunnel test data are taken from Ref.[37] The test model is a sharp-edged delta wing with aspect ratio $A = 2$, as shown in Fig. 4. The pitching axis is located at $67\%c_0$, where c_0 is the wing chord at midspan. The large-amplitude pitching oscillation tests were carried out in the 7 foot × 10 foot (1 foot = 30.48 cm) low-speed wind tunnel at NASA Ames Research Center, with Reynolds number $Re = 4.5 \times 10^5$ and the pitch moment reference point at $77\%c_0$.

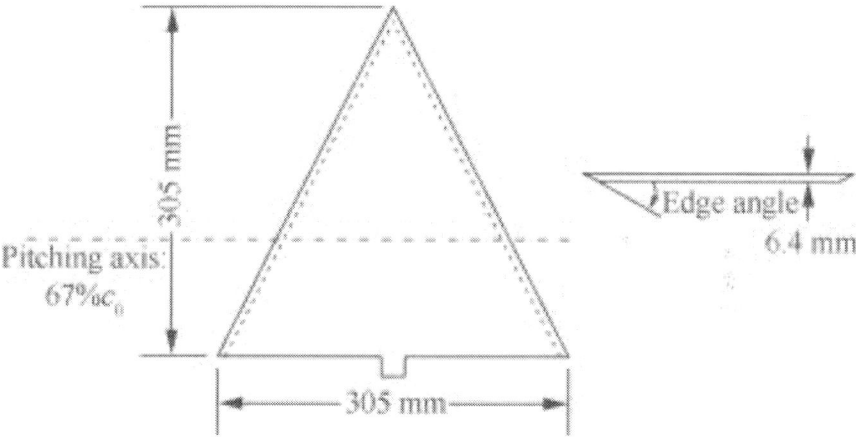

Figure 4. The delta wing model with aspect ratio $A = 2$.

In the forced-oscillations, the time history of angle of attack is

$$\alpha(\tau)=45°-45°\cos(k\tau) \tag{19}$$

where $k = \omega\bar{c}/V$ is reduced frequency, $\bar{c}$ is mean aerodynamic chord, ω is oscillation frequency, V is velocity; $\tau = t(V/\bar{c})$ is nondimensional time. The nondimensional parameter characterizing unsteadiness $K = 0.01, 0.02, 0.03$, and 0.04, where K is defined as

$$K = \dot{\alpha}_{\max}c_0/(2V) \tag{20}$$

Converted to the conventional reduced frequency, $k = 0.017, 0.034, 0.051, 0.068$.

The unsteady aerodynamic data are transformed to the dynamic increments:

$$\begin{cases} \Delta C_{L,\text{dyn}} = C_L - C_{L,\text{st}} \\ \Delta C_{D,\text{dyn}} = C_D - C_{D,\text{st}} \\ \Delta C_{m,\text{dyn}} = C_m - C_{m,\text{st}} \end{cases} \tag{21}$$

At first, the LS-SVM models of $\Delta C_{L,\text{dyn}}$, $\Delta C_{D,\text{dyn}}$, and $\Delta C_{m,\text{dyn}}$ are set up respectively from the dynamic measurement data with $K = 0.01, 0.03, 0.04$, with RBF kernel adopted, and the input variables taken as $\alpha\,(\tau)$, $\alpha\,(\tau - 8)$, $\alpha\,(\tau - 16)$, $\alpha\,(\tau - 24)$, $\alpha\,(\tau - 32)$, and $\bar{q}\,(\tau\ )$, where $\bar{q}$ is nondimensional pitching rate. This means that the instantaneous aerodynamics $C_m(\tau)$, for example, depends upon the angle-of-attack history during $[\tau - 32, \tau]$.

The SVM parameters determined by the m-fold cross-validation are listed as follows (with the sample inputs normalized):

$$C_D : \quad c = 3.2,\ 2\sigma^2 = 0.22$$
$$C_L : \quad c = 3.2,\ 2\sigma^2 = 0.21$$
$$C_m : \quad c = 3.2,\ 2\sigma^2 = 0.22$$

Fig. 5 shows the aerodynamic coefficients predicted by the LS-SVM models, in comparison with the measurement results, where the black dashed lines denote the static wind tunnel test data, the colored symbols are the dynamic wind tunnel test data, the colored lines are the predictions of the LS-SVM models, and the arrows indicate the direction of hysteretic loops. The predictions approximate the wind tunnel test data well, which indicates that the LS-SVMs have great learning ability.

Subsequently, the LS-SVM models are utilized to predict the aerodynamic characteristics of the large-amplitude pitching oscillation with $K = 0.02$. Fig. 6 shows the predicted aerodynamics in comparison with that obtained by wind tunnel tests. The predictions are in agreement with the test data, in spite of some tolerable discrepancies, which shows that the LS-SVMs have satisfying generalization performance.

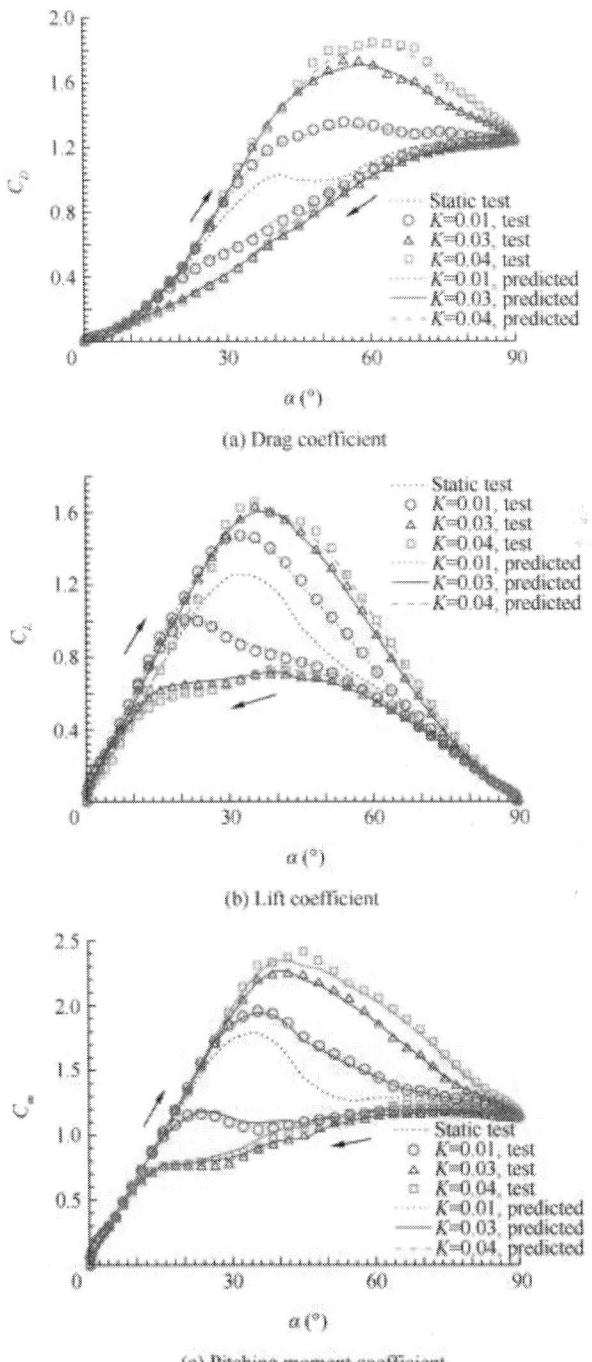

Figure 5. LS-SVM predictions and training data of pitching delta wing.

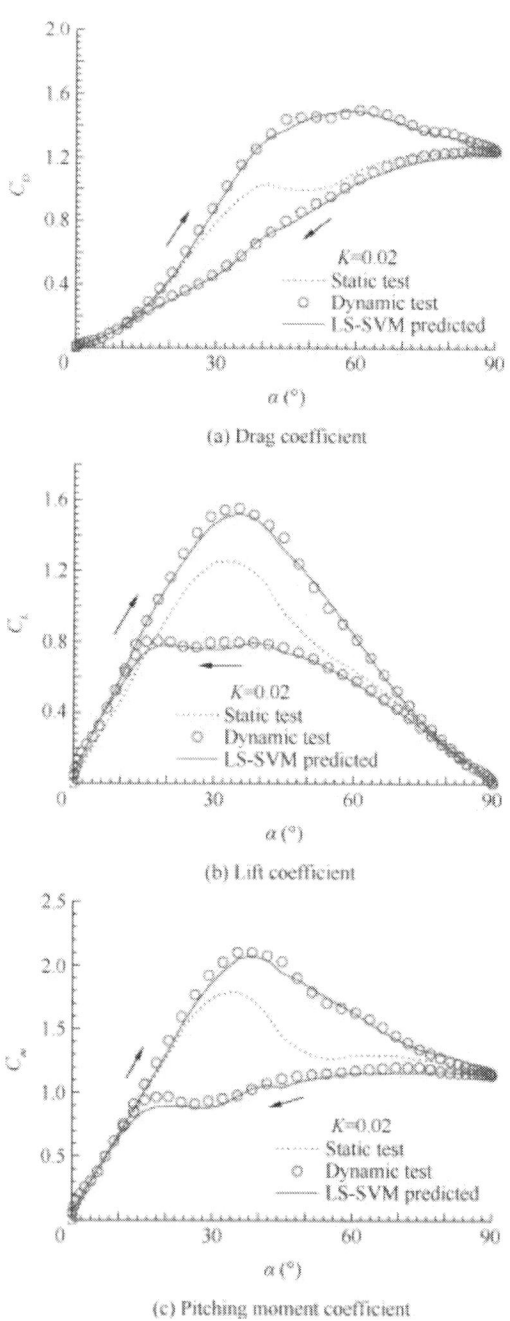

(a) Drag coefficient

(b) Lift coefficient

(c) Pitching moment coefficient

Figure 6. LS-SVM generalization and wind tunnel test data of pitching delta wing.

Aerodynamic modeling of a rolling aircraft configuration

Ref.[25] presented the rolling oscillation wind tunnel test results of F-16XL. The tests were executed with a 0.18-scale F-16XL model using a forced-oscillation rig in the 14 foot × 22 foot subsonic wind tunnel at NASA Langley Research Center. The test model was forced to roll around the longitudinal axis at the given pitch angles: $\phi = \phi_s \sin(k\tau$). Here, the reduced frequency $k=\omega b'/(2V)$, the nondimensional time $\tau=t(2V/b')$, b' is wing span.

In rolling oscillations, the angle of attack and sideslip angle vary as follows.

$$\begin{cases} \alpha(\tau) = \tan^{-1}(\tan\theta\cos\phi) \\ \beta(\tau) = \sin^{-1}(\sin\theta\sin\phi) \end{cases} \qquad (22)$$

where θ is the pitch angle and ϕ the roll angle.

The roll moment data were obtained at a series of pitching angles, $\theta = 0°$, 10°, 15°, 20°, 25°, 30°, 36°, 40°, 50°, 60°, 70°, and 75°, with the amplitude $\phi_s = 10°$, 20°, and 30°, and a constant maximum roll rate of $\bar{P}_{max} = P_{max}b'/(2V) = 0.04$.

The wind tunnel (W.T.) data used here are obtained from digitizing the AIAA paper. The error of roll moment coefficient C_l in the data points is less than 0.0005.

Aerodynamic modeling is performed directly for the roll moment coefficient due to lack of the corresponding static test data. At first, the test data with $\phi_s = 10°$ and 30° are employed to train the LS-SVM model of C_l, where RBF kernel is adopted also, and the input variables take $\alpha(\tau)$, $\alpha(\tau - 5)$, $\alpha(\tau - 10)$, $\alpha(\tau - 15)$, $\alpha(\tau - 20)$, $\beta(\tau)$, $\beta(\tau - 5)$, $\beta(\tau - 10)$, $\beta(\tau - 15)$, $\beta(\tau - 20)$, and $\bar{P}(\tau)$. The SVM parameters determined by m-fold cross-validation are (with the sample inputs normalized): $c = 5.0$ and $2\sigma^2 = 0.58$. The roll moment coefficient predicted by the LS-SVM model, as well as the one obtained from wind tunnel measurement, is presented in Fig. 7, where the red circles are the dynamic wind tunnel test data, the blue solid lines denote the predictions of the LS-SVM model, and the arrows indicate the direction of hysteretic loops. The figure shows a satisfactory approximation.

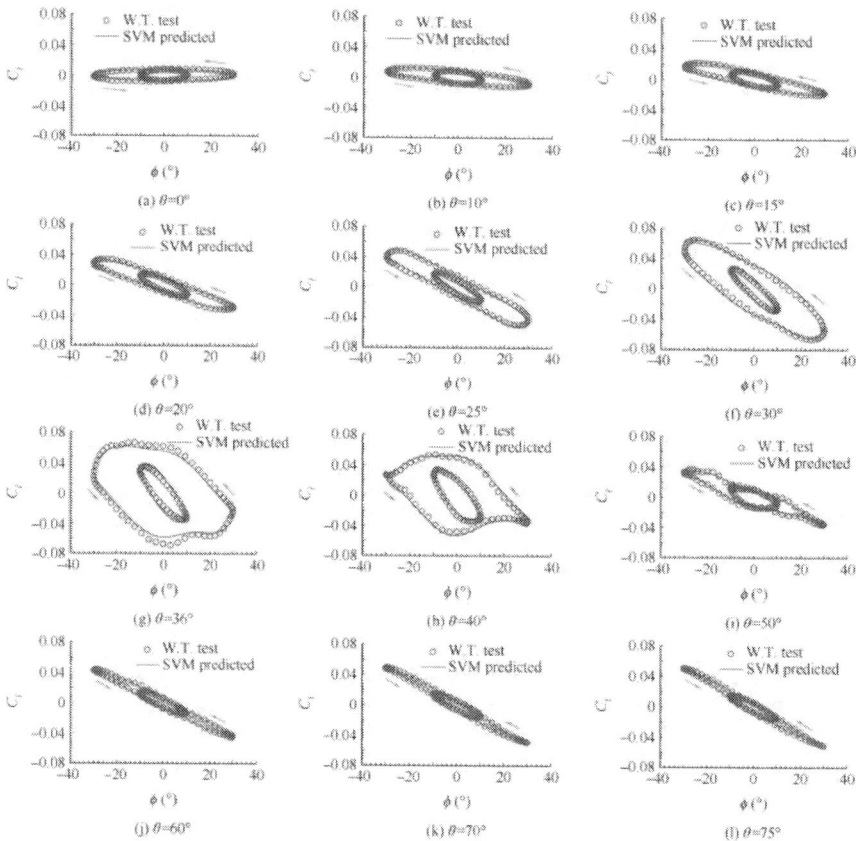

Figure 7. Prediction results of LS-SVM model and training data for rolling aircraft configuration.

The LS-SVM model is then employed to predict the aerodynamic response to the rolling oscillations with the amplitude $\phi_s = 20°$. Fig. 8 shows the predictions in comparison with the wind tunnel test data. It can be seen that there are some tolerable discrepancies in the range of $\alpha_0 = 30°–50°$, where the unsteady aerodynamic effects are extremely great. Nevertheless, the overall agreement indicates the satisfying generalization performance of LS-SVMs.

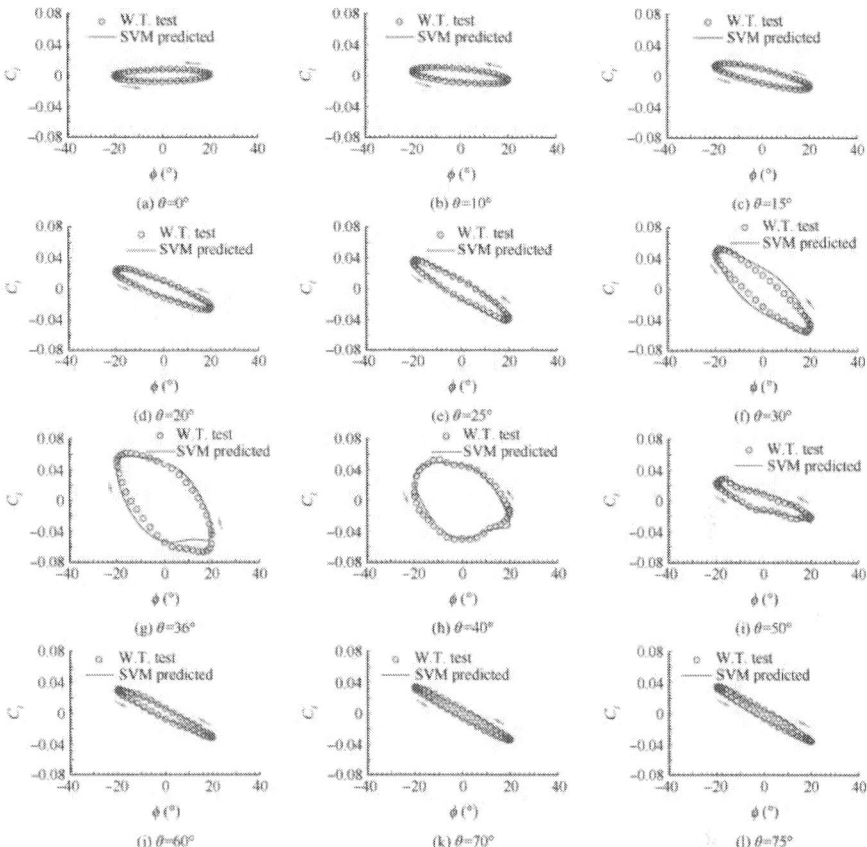

Figure 8. Generalization results of LS-SVM model and wind tunnel data for rolling aircraft configuration.

CONCLUSIONS

SVMs are a novel type of machine learning method developed on the basis of statistical learning theory, embodying the structural risk minimization principle. They are gaining popularity due to the features such as simple structure, global optimality and empirical performance. The following can be concluded from the presentations and applications in this paper.

(1) The aerodynamic modeling results of the pitching delta wing and the rolling aircraft configuration show that the LS-SVMs have excellent learning capability and satisfying generalization performance, and thus

become an attractive means in the field of high angle-of-attack unsteady aerodynamic modeling.

(2) In order to describe the effects of motion history on aerodynamics, one can take several sampling points of the current and previous flight states as the inputs of SVMs. It is suggested to model the dynamic increments instead of aerodynamic coefficients, for the static wind tunnel test data have higher precision than dynamic ones in general.

(3) RBF kernel is an appropriate selection for unsteady aerodynamic modeling at high angles of attack. The penalty factor c and the kernel width σ can be determined by means of m-fold cross-validation. It is important that all the data of an individual dynamic test should be taken as a unit and put into same a subset during the m-fold partitioning.

ACKNOWLEDGMENTS

The authors are grateful to all the members of their work group for help in the present research. They also thank the anonymous reviewers for their constructive review of the manuscript. This study was supported by a Chinese Government Contract.

REFERENCES

1. Lin GF, Lan CE. A generalized dynamic aerodynamic coefficient model for flight dynamics applications. Reston: AIAA; 1997. Report No.: AIAA-1997-3643.

2. Tobak M, Schiff LB. Aerodynamic mathematical modeling-basic concepts. Moffett Field, CA:NASA Ames Research Center; 1981. Report No.: 19810022564.

3. Chin S, Lan CE. Fourier functional analysis for unsteady aerodynamic modeling. AIAA J 1992;30(9):2259–66.

4. Murphy PC, Klein V. Estimation of aircraft nonlinear unsteady parameters from dynamic wind tunnel testing. Reston: AIAA; 2001. Report No.: AIAA-2001-4016.

5. Goman M, Khrabrov A. State-space representation of aerodynamic characteristics of an aircraft at high angles of attack. J Aircraft 1994;31(5):1109–15.

6. Lutze FH, Fan YG, Stagg G. Multi-axis unsteady aerodynamic characteristics of an aircraft. Reston: AIAA; 1999. Report No.: AIAA-1999-4011.

7. Abramov NB, Goman MG, Khrabrov AN. Aircraft dynamics at high incidence flight with account of unsteady aerodynamic effects. Reston: AIAA; 2004. Report No.: AIAA-2004-5274.

8. Abramov N, Goman M, Demenkov M, Khrabrov A. Lateraldirectional aircraft dynamics at high incidence flight with account of unsteady aerodynamic effects. Reston: AIAA; 2005. Report No.: AIAA-2005-6331.

9. Wang Q, Cai JS. Unsteady aerodynamic modeling and identification of airplane at high angles of attack. Acta Aeronaut Astronaut Sin 1996;17(4):391–8 Chinese.

10. Wang Q, He KF, Qian W, Mao ZJ. Aerodynamic modeling of spatial maneuvering aircraft at high angles of attack. Acta Aeronaut Astronaut Sin 2004;25(5):447–51 Chinese.

11. Huang XZ. Nonlinear indicial response/internal state-space representation and its application on delta wing aerodynamics. Reston: AIAA; 2003. Report No.: AIAA-2003-3944.

12. Pashilkar AA. Flight dynamic analysis of the flow incidence rate model. Reston: AIAA; 2002. Report No.: AIAA-2002-0098.

13. Wang Z, Lan CE, Brandon JM. Fuzzy logic modeling of nonlinear unsteady aerodynamics. Reston: AIAA; 1998. Report No.: AIAA- 1998-4351.

14. Wang Z, Lan CE, Brandon JM. Fuzzy logic modeling of lateraldirectional unsteady aerodynamics. Reston: AIAA; 1999. Report No.: AIAA-1999-4012.

15. Wang Z, Lan CE. Unsteady aerodynamic effects on the flight characteristics of an F-16XL configuration. Reston: AIAA; 2000. Report No.: AIAA-2000-3901.

16. Rokhsaz K, Steck JE. Application of artificial neural networks in nonlinear aerodynamics and aircraft design. J Aerospace 1993;102:1790–8.

17. Soltani MR, Sadati N, Davari AR. Neural network: a new prediction tool for estimating the aerodynamic behavior of a pitching delta wing. Reston: AIAA; 2003. Report No.: AIAA- 2003-3793.

18. Shi ZW, Wang ZH, Li JC. The research of RBFNN in modeling of nonlinear unsteady aerodynamics. Acta Aerodyn Sin 2012;30(1): 108–12 Chinese.

19. Wang Q, He KF, Qian WQ, Zhang TJ, Cheng YQ. Unsteady aerodynamics modeling for flight dynamics application. Acta Mech Sin 2012;28(1):14–23.

20. Wang Q, Wu KY, Zhang TJ, Kong YN, Qian WQ. Aerodynamic modeling and parameter estimation from the QAR data of an airplane approaching a high-altitude airport. Chin J Aeronaut 2012;25(3):361–71.

21. Singh R, Baeder J. Direct calculation of three-dimensional indicial lift response using computational fluid dynamics. J Aircraft 1997;34(4):465–71.

22. Ghoreyshi M, Cummings RM, DaRonch A, Badcock KJ. Transonic aerodynamic load modeling of X-31 aircraft pitching motions. AIAA J 2013;51(10):2447–64.

23. Ghoreyshi M, Jirasek A, Cummings RM. Reduced order unsteady aerodynamic modeling for stability and control analysis using computational fluid dynamics. Prog Aerosp Sci 2014;71:167–217.

24. Cooperative programme on dynamic wind tunnel experiment for manoeuvering aircraft. Reston: AIAA; 1996. Report No.: AGARD-AR-305.

25. Brandon JM, Foster JV. Recent dynamic measurements and considerations for aerodynamic modeling of fighter airplane configurations. Reston: AIAA; 1998. Report No.: AIAA-1998- 4447.

26. Huang D. Unsteady aerodynamic characteristics for the aircraft oscillation in large amplitude [dissertation]. Nanjing: Nanjing University of Aeronautics and Astronautics; 2007 [Chinese].

27. Findlay D, Guruswamy G. Numerical analysis of aircraft high angle of attack unsteady flows. Reston: AIAA; 2000. Report No.: AIAA-2000-1946.

28. Schu¨tte A, Einarsson G, Raichle A, Schoning B, Mo¨nnich W, Forkert T. Numerical simulation of manoeuvreing aircraft by aerodynamic, flight mechanics, and structural mechanics coupling. J Aircraft 2009;46(1):53–64.

29. Paul R, Murua J, Gopalarathnam A. Unsteady and post-stall aerodynamic modeling for flight dynamics simulation. Reston: AIAA; 2014. Report No.: AIAA-2014-0729.

30. Gunn SR. Support vector machines for classification and regression. Southampton (UK): University of Southampton; 1998. Report No.: ISIS-1-98.

31. Scho¨lkopf B, Sung K-K, Burges CJC, Girosi F, Niyogi P, Poggio T, et al. Comparing support vector machines with Gaussian kernels to radial basis function classifiers. IEEE Trans Signal Process 1997;45(11):2758–65.

32. Fan H-Y, Dulikravich GS, Han Z-X. Aerodynamic data modeling using support vector machines. Reston: AIAA; 2004. Report No.: AIAA-2004-0280.

33. Vapnik V. The nature of statistical learning theory. New York: Springer-Verlag; 1995.

34. Suykens JAK, Vandewalle J. Least squares support vector machine classifiers. Neural Processing Letters 1999;9(3):293–300.

35. Chapelle O, Vapnik V, Bousqet O, Mukherjee S. Choosing multiple parameters support vector machines. Machine Learning 2002;46(1):131–60.

36. Deng NY, Tian YJ. A new method for data mining: support vector machines. Beijing: Science Press; 2004, p. 355-66 Chinese.

37. Jarrah MA, Ashley H. Impact of unsteadiness on maneuvers and loads of agile aircraft. Reston: AIAA; 1989. Report No.: AIAA- 1989-1282-CP.

CITATION

Qing Wang, Weiqi Qian, Kaifeng He, Unsteady aerodynamic modeling at high angles of attack using support vector machines, Chinese Journal of Aeronautics, Volume 28, Issue 3, June 2015, Pages 659-668, ISSN 1000-9361, http://dx.doi.org/10.1016/j.cja.2015.03.010.

CHAPTER 10

Uncertainty Analysis and Design Optimization of Hybrid Rocket Motor Powered Vehicle for Suborbital Flight

Hao Zhu[1], Hui Tian[1], Guobiao Cai[1], Weimin Bao[2]

[1] School of Astronautics, Beihang University, Beijing 100191, China
[2] China Aerospace Science and Technology Corporation, Beijing 100048, China

ABSTRACT

In this paper, we propose an uncertainty analysis and design optimization method and its applications on a hybrid rocket motor (HRM) powered vehicle. The multidisciplinary design model of the rocket system is established and the design uncertainties are quantified. The sensitivity analysis of the uncertainties shows that the uncertainty generated from the error of fuel regression rate model has the most significant effect on the system performances. Then the differences between deterministic design optimization (DDO) and uncertainty-based design optimization (UDO) are discussed. Two newly formed uncertainty analysis methods, including the Kriging-based Monte Carlo simulation (KMCS) and Kriging-based Taylor series approximation (KTSA), are carried out using a global approximation Kriging modeling method. Based on the system design model and the results of design uncertainty analysis, the design optimization of an HRM powered vehicle for suborbital flight is implemented using three design optimization methods: DDO, KMCS and KTSA. The comparisons indicate that the two UDO methods can enhance the design reliability and robustness. The researches and methods proposed in this paper can provide a better way for the general design of HRM powered vehicles.

INTRODUCTION

With the increasing demands for green, nontoxic and cheap propulsion technologies, hybrid rocket motors (HRMs) show great potential as they are less complex and cheaper than liquid rocket motors (LRMs), and more easily throttled and restarted than solid rocket motors (SRMs).[1, 2 and 3] It makes sense to develop sub-orbit vehicles with HRMs which have such advantages as safety, cheapness and non-toxicity, since the near space of 30–100 km altitude is becoming increasingly important in scientific research and military applications in recent years. Therefore, there are many academic studies and projects about sub-orbit vehicles with HRMs recently.[4, 5, 6 and 7]

It is necessary and important to apply design optimization methods in the aerospace vehicle design process in order to improve the design level and efficiency. In the traditional design optimization methods, the input parameters are considered as deterministic values to simplify the modeling process. However, it may be inconsistent with the objective reality. Therefore, the studies on the uncertainties in the aerospace vehicle design process have important theoretical and practical values to improve the overall design level.

Compared with the traditional SRMs or LRMs, HRMs have both a liquid oxidizer feeding system and a solid fuel combustion chamber, so the system design model of HRMs has more input variables and model parameters. Moreover, since the combustion mechanism of HRMs is not fully researched at present, there are more uncertainties in the design process of HRMs. The uncertainties probably result in the fact that the optimal design results under deterministic design optimization (DDO) are infeasible or unreliable in the following manufacturing process. Nevertheless, the current studies on design optimization of HRMs or its applications typically focus on the DDO method,[4 and 5] so it is necessary to study the uncertainties and develop uncertainty-based design optimization (UDO) methods to enhance the design reliability and robustness. Therefore, an approach to the uncertainty analysis and design optimization of HRM powered vehicles is proposed in this paper, based on our former work about the conceptual design of HRM powered rockets.[8 and 9]

The main problem in UDO is the low efficiency of the uncertainty analysis when the system design model is complicated. The approximate model technology is one of the most popular methods to solve this problem. Kriging model is a widely used approximate model for its advantages such as unbiased estimator at the training sample point, desirably strong nonlinear approximating ability, flexible parameter selection of the model and accurate global approximation ability.10 [and] 11 An approach that applying the Kriging model to two uncertainty analysis methods, i.e., Monte Carlo simulation (MCS) and Taylor series approximation (TSA), is proposed in this paper. Both newly formed methods are applied to the design optimization of the HRM powered vehicle for suborbital flight and the design results with high reliability and robustness are obtained.

SYSTEM DESIGN MODEL

The HRM powered sub-orbit vehicle is a ballistic rocket with an aerodynamic stable shape. The system design process involves many disciplines including structure, propulsion, aerodynamic, launching dynamics and trajectory. Each discipline is analyzed to find out possible mathematical relationships between design variables and performance parameters, such as the rocket lift-off mass M_R or the rocket body lengthL_R, and develop a feasible multidisciplinary design model of the rocket system.

ROCKET STRUCTURE DESIGN

The structure of the HRM powered rocket consists of head (containing payloads), fins, HRM and the linking structures,[8] as shown in Fig. 1. The rocket lift-off mass M_R can be obtained by

$$M_R = M_m + M_s + M_{pay} \tag{1}$$

where M_{pay} is the payload mass. The HRM mass M_m is deduced by HRM design. The rocket structure mass M_s consists of head mass, fin mass and linking structures mass. It is related to rocket diameter D_R and defined as

$$M_s = 75D_R - 7.5 \tag{2}$$

The rocket body length L_R can be obtained by

$$L_R = L_m + L_h \tag{3}$$

where L_h is the rocket head length and it is 1 m in this paper. The HRM length L_m is also deduced by HRM design.

where L_h is the rocket head length and it is 1 m in this paper. The HRM length L_m is also deduced by HRM design.

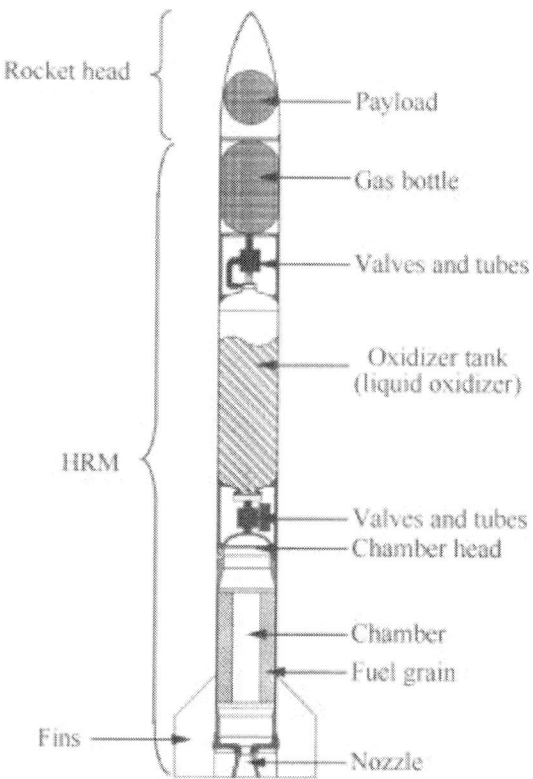

Figure 1. HRM powered rocket structure.

HRM design

HRM is the main part of the rocket. Its mass and dimension almost determine the mass, dimension and trajectory of the rocket. A wheel port grain is selected in the HRM with a propellant combination of 98% hydrogen peroxide (HP) and hydroxyl-terminated polybutadiene (HTPB). A nitrogen gas pressure feed subsystem is used. The oxidizer mass flow rate is controlled to keep constant by an ideal venturi section. The HRM is designed as shown in Ref.[8] The propellant, including the solid fuel and liquid oxidizer, constitutes the main mass and dimension of the HRM, therefore it determines M_R and L_R indirectly.

Launch simulation

A 35 m length ramp is used in the launch subsystem. The dynamic equation of the ramp launch process can be obtained based on Newton's Second Law as follows:

$$\dot{V} = F/M - g \sin \varphi_0 - \lambda g \cos \varphi_0 \qquad (4)$$

where V is the rocket velocity, F the thrust, M the rocket mass that changes as the HRM combusts, g gravity constant and φ_0 the launch elevation angle which needs to be optimized. The friction coefficient λ is 0.05. The aerodynamics drag is ignored because the rocket velocity is slow and the launch phase time is short.

Trajectory simulation

A two degree of freedom (DOF) mass point trajectory equation is used with no control law as follows:

$$\ddot{X} = (F - C_D q S_M)\dot{X}/V/M \qquad (5)$$

$$\ddot{Y} = (F - C_D q S_M)\dot{Y}/V/M - g \qquad (6)$$

where X and Y are the horizontal and vertical distances of the position from the launch point, q is the dynamic pressure and C_D represents the drag coefficient. Rocket body section area is used as the reference area S_M.

Dynamic computation

A U.S. 76 standard atmosphere model with no atmosphere motion is used here. The drag coefficient C_D of "Titan II" rocket [12] is used as shown in Table 1, given a 20% increase as the sub-orbit rocket in this paper has fins.

Table 1. Drag coefficient of USA "Titan II rocket". [12]

Ma	C_D
$0 \leqslant Ma < 0.8$	0.29
$0.8 \leqslant Ma < 1.068$	$Ma - 0.51$
$Ma \geqslant 1.068$	$0.091 + 0.5\, Ma^{-1}$

Based on the analysis and deduction above, the multidisciplinary design model of the rocket system is established. There are 43 input parameters, including 8 design variables (shown in Table 2) and 35 model parameters (shown in Table 3), in the mathematical model through which the performance response parameters, such as M_R and L_R, are computed.

Table 2. Design variables.

Design variable	Lower limit	Upper limit	Distribution	Relative limit deviation
Grain outer diameter D_p (m)	0.2	0.5	Normal	0.005
Initial web thickness e_1 (m)	0.02	0.05	Normal	0.01
Number of wheel port n	3	10	Not uncertain factor	
Initial thrust F_i (kN)	10	20	Uniform	0.05
Initial chamber pressure p_{ci}(MPa)	1	3	Uniform	0.05
Initial oxidizer to fuel ratio α_i	2	5	Uniform	0.05
Nozzle expansion ratio ε	3	10	Normal	0.015
Launch elevation angle φ_0 (°)	65	85	Normal	0.005

Table 3. Model parameters.

Model parameter	Symbol	Mean	Limit deviation	Distribution
Regression rate precision coefficient	x1	1	0.02	Uniform
Weld coefficient	x2	0.8	0.03	Uniform
Pressure oscillation coefficient	x3	1.2	0.03	Uniform
Volume fraction of remained oxidizer	x4	0.05	0.05	Uniform
Volume fraction of initial air cushion	x5	0.05	0.05	Uniform
Combustion efficiency	x6	0.96	0.015	Uniform
Nozzle efficiency	x7	0.93	0.015	Uniform
Initial tank temperature (K)	x8	293.15	0.0682	Uniform
Injector pressure drop coefficient	x9	0.2	0.0325	Normal
Tube pressure loss coefficient	x10	0.2	0.08	Normal
Initial gas bottle pressure (MPa)	x11	30	0.03	Normal
Fuel density (kg/m³)	x12	1218	0.015	Normal
Oxidizer density (kg/m³)	x13	1440	0.015	Normal
Chamber heat insulation layer density (kg/m³)	x14	1000	0.015	Normal
Chamber and nozzle shell density (kg/m³)	x15	7750	0.015	Normal
Chamber head density (kg/m³)	x16	7750	0.015	Normal
Tank shell density (kg/m³)	x17	2850	0.015	Normal
Gas bottle shell density (kg/m³)	x18	1750	0.015	Normal

Semi-minor axis length ratio of chamber head	x19	3	0.01	Normal
Semi-minor axis length ratio of oxidizer tank head	x20	2	0.01	Normal
Semi-minor axis length ratio of gas bottle head	x21	2	0.01	Normal
Chamber heat insulation layer thickness (m)	x22	0.003	0.025	Normal
Nozzle heat insulation layer thickness (m)	x23	0.015	0.025	Normal
Injector panel thickness (m)	x24	0.005	0.02	Normal
Minimum machining thickness of chamber and nozzle shell material (m)	x25	0.0015	0.066	Normal
Minimum machining thickness of oxidizer tank shell material (m)	x26	0.0015	0.066	Normal
Minimum machining thickness of gas bottle shell material (m)	x27	0.005	0.025	Normal
Nozzle half expansion angle (°)	x28	15	0.012	Normal
Nozzle half convergence angle (°)	x29	45	0.012	Normal
Yield limit of chamber and nozzle shell material (MPa)	x30	1080	0.01	Normal
Yield limit of chamber head material (MPa)	x31	1080	0.01	Normal
Yield limit of oxidizer tank shell material (MPa)	x32	490	0.01	Normal
Yield limit of gas bottle shell material (MPa)	x33	1760	0.01	Normal
Ramp length (m)	x34	35	0.0001	Normal
Drag precision coefficient	x35	1	0.1	Normal

DESIGN UNCERTAINTY ANALYSIS

Design uncertainties

In design phase, specifically the simulation-based computational design, there are three sources contributing to the total uncertainty of computational simulation, namely model input uncertainty, model uncertainty, and model error.[13] The model input uncertainties constitute the main part of the total uncertainties, so only they are considered in this paper. The model input uncertainties are the uncertainty factors in the design variables and model parameter. Most of them are emerged from the engineering realization process. Except for the number of wheel port n, all the other 42 input parameters are uncertain factors in the hybrid rocket system design model. The most accurate methodology to classify and quantify these uncertain factors is probability theory, which is applied in this study. The uncertainties in the three types of the input parameters, including physical or chemistry characteristic parameters such as fuel density, machining parameters such as grain outer diameter and parameters obtained by testing(such as injector pressure drop coefficient), are classified to be normal distribution. The uncertainties in the other parameters, which are either generated from the cognitive incompletion of the physical word such as the regression rate, or determined by these parameters directly or indirectly such as the initial thrust, are classified to be uniform distribution. According to design criteria and engineering experience 14, 15 and 16, the relative limit deviations of the design uncertainties are confirmed as shown in Table 2 and Table 3. The standard deviations of uncertainties with normal distribution can be evaluated in terms of "6σ" principle.

At each step of the optimization process, the uncertainties of design variables can be quantified as

$$\begin{cases} x'_L = -x'\varDelta \\ x'_U = x'\varDelta \end{cases}$$

$$(7)$$

where $\varDelta$ is the relative limit deviation; x'_L and x'_U are the upper limit deviation and the lower limit deviation of the design variable when its value is x'.

Sensitivity analysis

It makes the uncertainty analysis complicated and time-consuming with a lot of input uncertainties, so a sensitivity analysis method[17] is used to filter out the insignificant model parameter uncertainties in the UDO approach of hybrid rocket design. According to the probability distributions, the model parameter uncertainties are sampled 1000 times respectively with a Latin hypercube sampling (LHS) method.[18] Then the main performance parameters are computed by the system design model. The sensitivities of the rocket's main performance parameters, including M_R, the top altitude of trajectory Y_{max}, the maximum rocket axes acceleration N_{xmax}, and the rocket length to diameter ratio L/D are obtained by the sensitivity analysis method. The top ten model parameter uncertainties that have the most significant effects on M_R, Y_{max}, N_{xmax}, and L/D are shown respectively from Fig. 2, Fig. 3, Fig. 4 and Fig. 5. The length of the bar in the figures represents the sensitivity of the performance parameters to model parameter uncertainties. The positive value of bar means that the performance parameter increases when the model parameter uncertainty increases and vice versa.

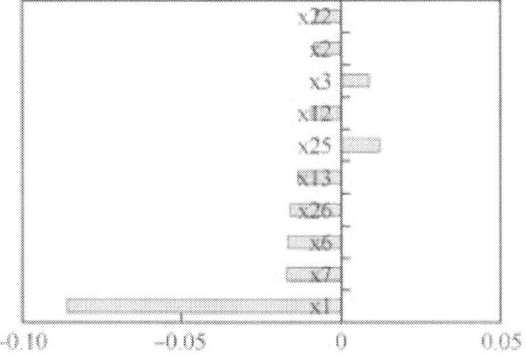

Figure 2. Sensitivity of M_R to model uncertain factors.

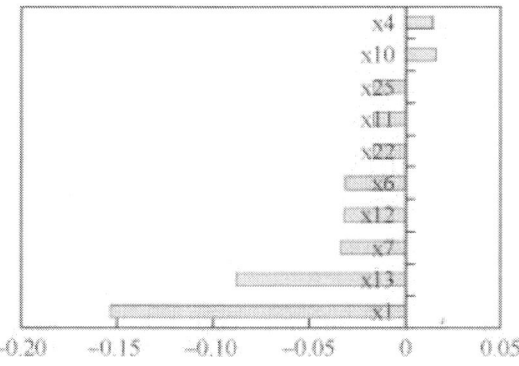

Figure 3. Sensitivity of L/D to model uncertain factors.

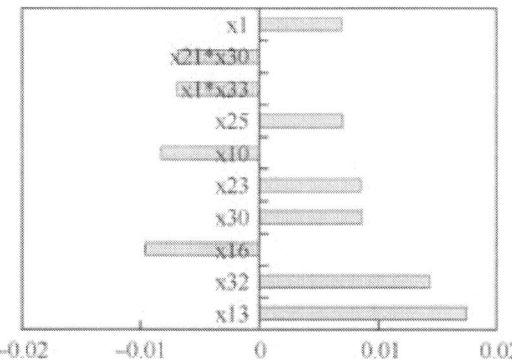

Figure 4. Sensitivity of N_{xmax} to model uncertain factors.

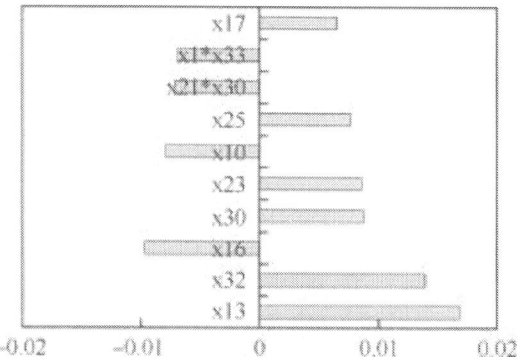

Figure 5. Sensitivity of Y_{max} to model uncertain factors.

When the oxidizer mass flow rate is constant, the regression rate determines the burning time, which directly determines the oxidizer mass M_o; M_o has an important influence on M_R and L_R as discussed in Section 2.2. Therefore, the fuel regression rate precision coefficient (x1) has the greatest impact on M_R and L/D indirectly as shown in Fig. 2 and Fig. 3. The oxidizer density determines the oxidizer mass which influences the rocket structure mass. The yield limit of oxidizer tank shell material determines the tank wall thickness which influences the tank mass M_t consequently. In addition, N_{xmax} and Y_{max} are mainly determined by the rocket velocity, attitude and altitude at the burnout point which is influenced by rocket structure mass according to Tsiolkovski formula. Therefore oxidizer density (x13) and the yield limit of oxidizer tank shell material (x32) have the greatest impact on N_{xmax} and Y_{max} indirectly as shown in Fig. 4 and Fig. 5. At the current level of science and technology, the uncertainty of material density and strength can be quantified accurately, while the fuel regression rate model of HRM is not investigated clearly enough. Most uncertain data of regression rate is mainly obtained through certain test of a special HRM and it is not suitable for all HRMs. Therefore uncertain factors in the regression rate are the main source of the rocket performance uncertainties using an HRM.

Through the above sensitivity analysis, three model parameter uncertainties, including the fuel regression rate precision coefficient (x1), oxidizer density (x13) and the yield limit of oxidizer tank shell material (x32) (shown in Table 3 with boldfaced words), together with 7 design variables (except for n) are selected as the uncertain factors in the following UDO approach.

KRIGING-BASED UDO

The Kriging-based uncertainty analysis methods are proposed in this section. The general UDO process and two uncertainty analysis method including MCS and TSA are discussed, then a global approximation Kriging modeling approach is presented. Two newly formed uncertainty analysis methods including the Kriging-based Monte Carlo simulation (KMCS) and Kriging-based Taylor series approximation (KTSA) are proposed sequentially.

UDO

The tradition optimization can be expressed in a mathematics model as

$$\begin{cases} \text{find} & x \\ \text{min} & f(x,p) \\ \text{s.t.} & g(x,p) \leqslant 0 \\ & x^L \leqslant x \leqslant x^U \end{cases} \tag{8}$$

where x is the design variables and p the model parameters, both of them are input parameters; x^L and x^U are the lower and upper boundaries of x, $f(x,p)$ is the objective function and $g(x,p)$ represents the constraint vector. In DDO, all of $x;p, f(x,p)$ and $g(x,p)$ are treated as deterministic parameters, and the optimal organization and the search strategy are based on the deterministic relationships as a result. The benefits are the simplification of the optimal process and the saving of computing time.

However, there are many uncertainties in the real world. In UDO, all design variables, model parameters and mathematical models are analyzed and the uncertain factors between them are separated. The uncertainties of the performance responses can be computed with different methods sequentially. The mathematical model of UDO can be expressed as

$$\begin{cases} \text{find} & x \\ \text{min} & F[\mu_f(x,p), \sigma_f(x,p)] \\ \text{s.t.} & P[g(x,p) \leqslant 0] \geqslant R \\ & x^L \leqslant x \leqslant x^U \end{cases} \tag{9}$$

where both x and p could be uncertain; μ_f and σ_f are the mean and standard deviation of the original optimization objective function f, and F is the reformulated optimization objective function with respect to μ_f and σ_f; P is the probability of the statement within the braces to be true, and R is the reliability vector specified for the constraint vector. The robustness of the system is achieved through minimizing μ_f and σ_f, and the reliability of constraints is accomplished through confidence level at which constraints are met with a higher probability.

Uncertainty analysis is a key procedure of UDO process. At this step, the uncertainty characteristics of the system responses propagated from the input uncertainties are quantitatively analyzed. There are many uncertainty analysis methods, including MCS, TSA, reliability analysis, etc.

MCS method is the most accurate solution for uncertainty problems based on probability theory. The mean value, standard deviation, distribution function and probability density of the responses are predicted statistically from the random sampling metric in MCS analysis. μ_f and σ_f are obtained as

$$\mu_f = \frac{\sum_{i=1}^{N} f(x_i)}{N} \tag{10}$$

$$\sigma_f = \frac{\sum_{i=1}^{N} (f(x_i) - \mu_f)}{N - 1} \tag{11}$$

where $f(x_i)$ is the response function about the x_i sample point. The prediction accuracy of MCS analysis is inversely proportional to the square root of the sampling number N, thus extensive sample number is needed to ensure the prediction accuracy. As a result, it is unacceptable for the optimization problems with a long-running time simulation program.

TSA method is one of the most efficient solutions for uncertainty problems with probability theory. Based on the first-order Taylor series, μ_f can be estimated as [13]

$$\mu_f = E(f(x)) \approx f(\mu_x) + \sum_{i=1}^{n} \frac{\partial f}{\partial x_i} E(x_i - \mu_{ix}) = f(\mu_x) \tag{12}$$

where μ_x is the mean value of n-dimensional vector x. If the input variables are not related, σ_f can be estimated as [13]

$$\sigma_f = \sqrt{\sum_{i=1}^{n} \left(\frac{\partial f}{\partial x_i}\right)^2 \sigma_{x_i}^2} \tag{13}$$

where σ_{xi} is the standard deviation of x_i. The efficiency of TSA method is much higher than that of the MCS method, since it does not need repeated calculation. However, when $f(x)$ is a nonlinear system whose first-order gradient cannot be obtained with analytical method, the application of TSA is restricted. How to obtain an accurate gradient value with low time cost is the key point to solve this problem.

Global approximation Kriging modeling

The Kriging model is an interpolation technique based on statistical theory. Its advantages, such as unbiased estimator at the training sample point, desirably strong nonlinear approximating ability, flexible parameter selection of the model and accurate global approximation ability, make it widely used in approximate models. It takes full account of the relevant characteristics of the variable space, containing the regression part and the nonparametric part

$$\hat{y}(x) = f(x) + z(x) \tag{14}$$

where $f(x)$ is deterministic function that is a global approximation of the design space represented by the polynomial of x; $z(x)$ is a Gaussian stochastic process with zero mean and variance. Ref. [10] showed the detailed developing process of Kriging model with N sampling points. When using Gauss function as the correlation function in Kriging model, the first-order derivative at point x_i can be estimated by

$$\frac{\partial \hat{y}(x)}{\partial x_i} = \frac{\partial f(x)}{\partial x_i} + \frac{\partial z(x)}{\partial x_i} \tag{15}$$

The approximation accuracy of the approximate model is closely related to the quantity of sample points, thus a global approximation Kriging modeling method is developed by sequentially sampling the design space and update the Kriging model in order to get higher approximation accuracy with fewer sample points. The process is shown in Fig. 6and the main steps are shown as follows:

Step 1. Use LHS method to generate the initial training points with a small scale.

Step 2. Establish the initial Kriging model.

Step 3. Use LHS method to generate the test points.

Step 4. Calculate the response values using the original model and the Kriging model respectively, then compare the results to get the approximation error of the Kriging model.

Step 5. According to the results of Step 4, judge if the error satisfies the precision criterion. The adjusted multiple correlation coefficient R_a^2 is used as the global precision criterion and the maximum relative error e_{max} is used as the local precision criterion. [19] If yes, end the iteration and output the present Kriging model; else, goto the next step.

Step 6. Choose the test points at which the relative errors are larger than e_{max} and put them into the training space to update the Kriging model in order to enhance its approximation accuracy at the design areas where the test points have large relative errors with the previous Kriging model.

Step 7. Return to Step 2, continue the iteration until the Kriging model satisfies the desired precision.

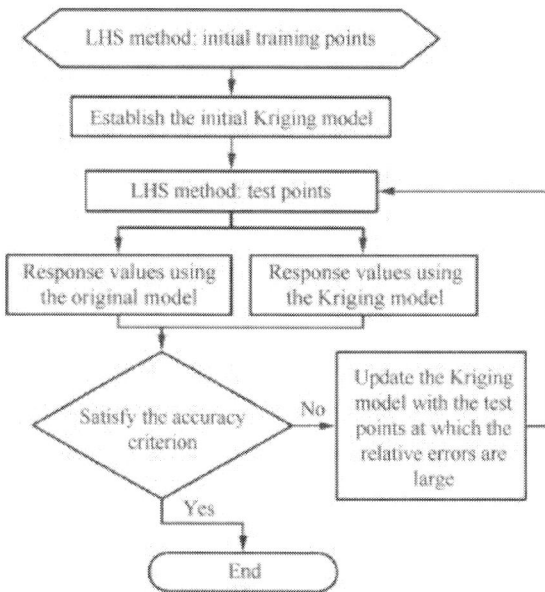

Figure 6. Procedures of global approximation Kriging modeling process.

There are many verifications about the approximation accuracy of Kriging model, so only the prediction precision of the first-order derivative is tested by a symmetric two-dimensional nonlinear function with multiple local extreme points[20] shown as

$$y(x_1, x_2) = 2 + 4x_1 + 4x_2 - x_1^2 - x_2^2 + 2\sin(2x_1)\sin(2x_2)$$
$$x_1, x_2 \in [0.5, 3.5]$$

(16)

The number of the initial training points is 10 to develop the Kriging model of the test function and the results are shown in Table 4. Along with the increase of the training point number m, R_a^2 increases to 1 gradually and e_{max} decreases. When the number of training points reaches 37, R_a^2 is close to 1 and e_{max} nearly decreases to 0, indicating that Kriging model has reached a good prediction precision of the first-order derivative as shown in Fig. 7 and Fig. 8.

Table 4. Iteration process of Kriging model for the test function.

Iteration	Training points	Test points	R_a^2	e_{max}
0	10	20	0.6934	0.3302
1	27	20	0.9832	0.0555
2	36	20	0.9961	0.0384
3	37	20	0.9995	0.0189

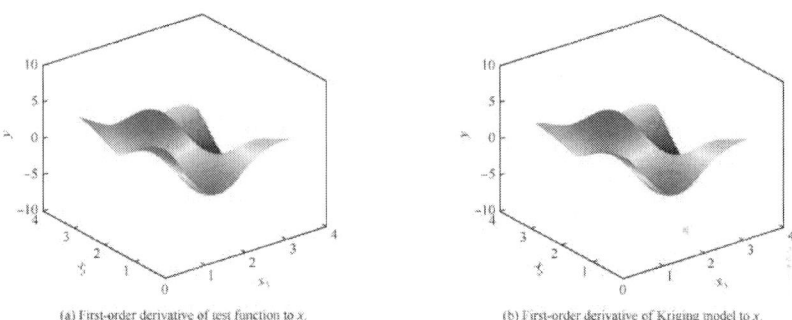

(a) First-order derivative of test function to x_1 (b) First-order derivative of Kriging model to x_1

Figure 7. First-order derivative of test function and its Kriging model to x_1.

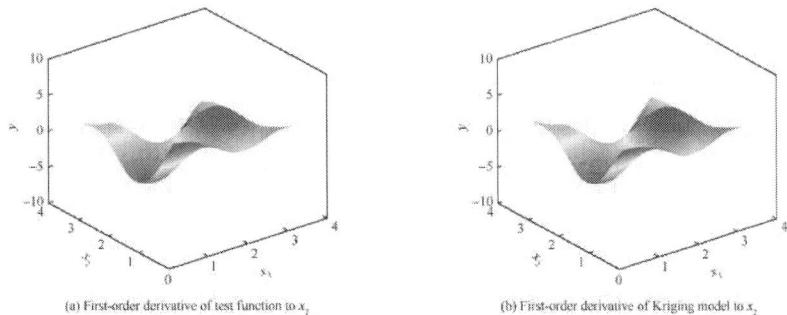

(a) First-order derivative of test function to x_2 (b) First-order derivative of Kriging model to x_2

Figure 8. First-order derivative of test function and its Kriging model to x_2.

Design optimization with KMCS

MCS is considered to be the most accurate method among the uncertainty analysis methods based on probability theory; however its computation precision is directly related to sampling frequency, thus a great number of sampling points are needed when MCS is used. It is unacceptable for the simulation-based design optimization problems whose system design model is complicated and time-consuming. Therefore the KMCS method is developed in which the original complicated system design model $f(x)$ used in MCS is surrogated by the approximate function $\hat{f}(x)$ established by the global approximation Kriging modeling method. The approximate mean value $\hat{\mu}_f$ and the standard deviation $\hat{\sigma}_f$ of the system responses in KMCS can be computed using Eqs.(17) and (18). Fig. 9 shows the procedure of UDO with KMCS method.

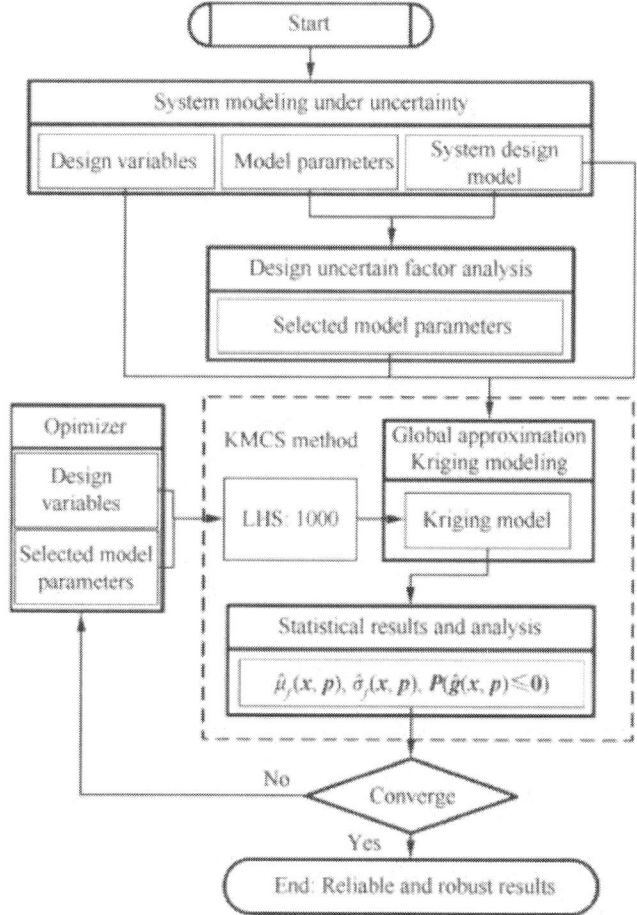

Figure 9. Procedures of UDO with KMCS method.

$$\hat{\mu}_f = \frac{\sum_{i=1}^{N} \hat{y}(x_i)}{N} \tag{17}$$

$$\hat{\sigma}_f = \frac{\sum_{i=1}^{N}\left(\hat{y}(x_i) - \hat{\mu}_f\right)}{N - 1} \tag{18}$$

Design optimization with KTSA

The Kriging model has a high prediction accuracy not only on the function value but also its first-order derivatives, thus the Kriging model is applied to obtaining the first-order gradient value of $f(x)$ used in the TSA method and it can effectively solve the problems that generated when TSA is applied to multidimensional nonlinear systems. With a Kriging model accurate enough, the KTSA method can effectively expand the application field of TSA method. The mean value and the standard deviation of the system responses can be computed using Eqs. (19) and (20), where $\frac{\partial \hat{y}(x)}{\partial x_i}$ is computed by Eq.(15). The "6σ" principle is used to compute constraint satisfaction probability as shown in Eqs. (21) and (22), where G is the reliability level, μ_g the mean of the constraint functions, σ_g the standard deviation of the constraint functions, and k Sigma level. For the normal distribution function, when $k = 6$, constraint satisfaction probability is 99.9999998%; when $k = 3$, constraint satisfaction probability is about 99.9937%. $k = 3$ is chosen in this paper. Fig. 10 shows the procedure of UDO with KTSA method.

$$\hat{\mu}_y = \hat{y}(\boldsymbol{\mu}_x) \tag{19}$$

$$\hat{\sigma}_y = \sqrt{\sum_{i=1}^{N}\left(\frac{\partial \hat{y}(x)}{\partial x_i}\right)^2 \sigma_{x_i}^2} \tag{20}$$

$$\mu_{g1}(\boldsymbol{x},\boldsymbol{p}) - k\sigma_{g1}(\boldsymbol{x},\boldsymbol{p}) \geqslant G_1 \tag{21}$$

$$\mu_{g2}(\boldsymbol{x},\boldsymbol{p}) + k\sigma_{g2}(\boldsymbol{x},\boldsymbol{p}) \leqslant G_2 \tag{22}$$

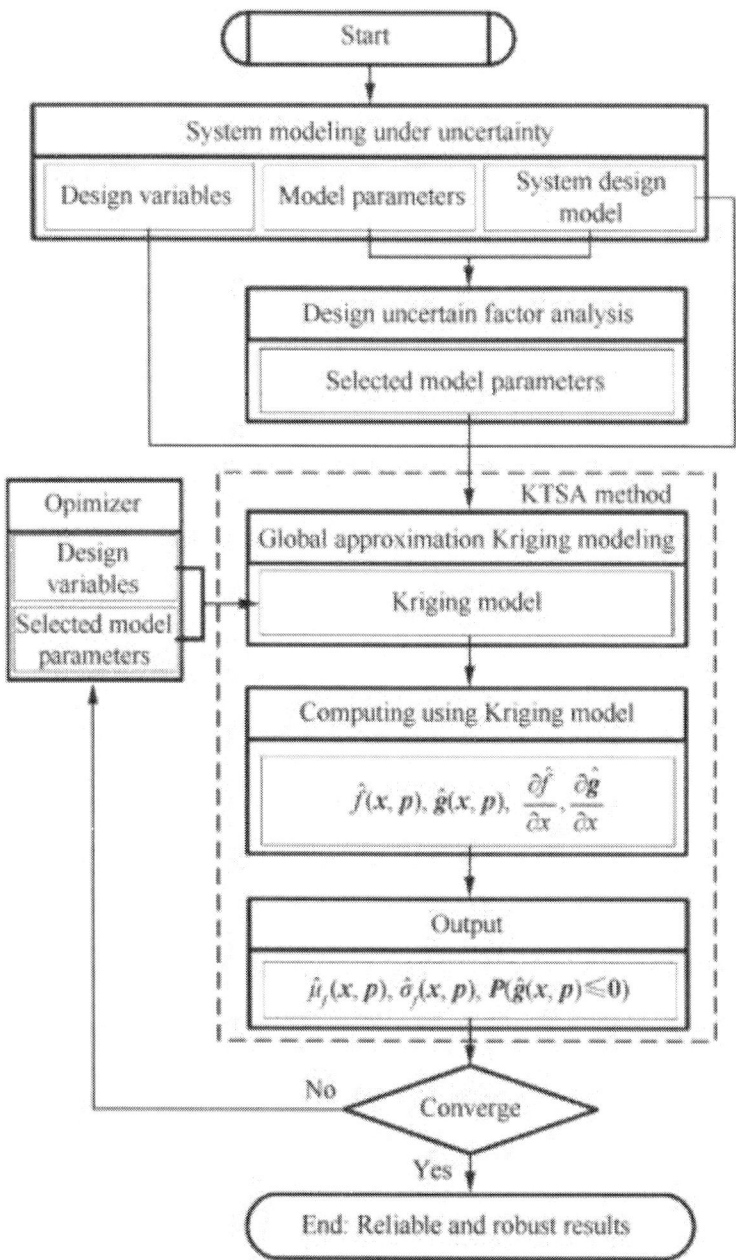

Figure 10. Procedure of UDO with KTSA method.

OPTIMIZATION, RESULTS AND DISCUSSION

Design variables, target function and constraints

The general mission of sub-orbit rocket is to detect the high-altitude environment or supply the payload with a certain microgravity time. A HRM powered rocket with a mission to send a 50 kg payload to an altitude over 120 km is designed in this study. The payloads are often avionic equipment which is sensitive to acceleration, so the maximum rocket axial acceleration N_{xmax} is constrained. One of the characteristics of the rocket body with a series-wound structure HRM is its great length, so a value of 18 is chosen to constrain the rocket length to diameter ratio L/D, considering that a too high L/D is not good for the rocket structure strength. Both the design difficulty and the cost of a rocket vehicle are mainly determined by M_R, so the target function is to minimize M_R by satisfying the constraints above. All the design variables x and their boundaries x^L, x^U are shown in Table 2 and the three model uncertain parameters are shown in Table 3 with boldfaced words. The mathematics model of the DDO is shown in Eq. (23).

$$
\begin{cases}
\text{find} & x \\
\text{min} & M_R = f(x,p) \\
\text{s.t.} & Y_{max} = g_1(x,p) \geqslant 120 \text{ km} \\
& L/D = g_2(x,p) \leqslant 18 \\
& N_{xmax} = g_3(x,p) \leqslant 10 \text{ g} \\
& x^L \leqslant x \leqslant x^U
\end{cases}
\tag{23}
$$

In UDO approach, the robustness of the system is achieved through minimizing the mean value μ_{MR} and standard deviation σ_{MR} of the target function while the reliability of constraints is accomplished through the confidence-level that the constraints are met with a higher probability, shown as Eq. (24).

$$
\begin{cases}
\text{find} & x \\
\text{min} & F[\mu_{M_R(x,p)}, \sigma_{M_R(x,p)}] \\
\text{s.t.} & P_1[Y_{max}(x,p) \geqslant 120km] \geqslant 0.95 \\
& P_2[L/D(x,p) \leqslant 18] \geqslant 0.95 \\
& P_3[N_{xmax}(x,p) \leqslant 10g] \geqslant 0.95 \\
& x^L \leqslant x \leqslant x^U
\end{cases}
\tag{24}
$$

Results and discussion

A modified differential evolution (MDE) algorithm[21] is applied to implementing global optimization and improving the efficiency and quality of the optimization solution, then DDO, KMCS and KTSA methods are applied respectively to the design optimization of the HRM powered rocket for a suborbital flight. The design variables n and φ_0 are set to be constant and equal to the DDO optimal result to simplify the calculation. The comparisons of the results are shown in Table 5, and the statistical results considering all the uncertain input parameters are shown in Table 6 and Fig. 11. Compared to the DDO results, the two UDO methods achieve reliability requirements at a higher confidence level and σ/μ of M_R reduces by 8.0% as shown in Table 6. The reason that reliability of N_{xmax} satisfying the constraint is 100% in DDO method is that optimal result is not at the boundary between feasible and unfeasible region as shown in Fig. 11. Although the mean value of the rocket lift-off mass is a little bigger than that of deterministic one, the reliability and robustness are enhanced obviously with two UDO method. Compared with KTSA, the results of KMCS are comparatively better while its efficiency is lower. The prediction precision of KTSA is not better than KMCS inherently since the former applies two approximate process, Kriging model and TSA method. As a result, the KTSA method is more suitable for the initial design optimization phase while the KMCS method is more applicable for the detailed design optimization phase. The results and comparisons prove that the uncertainty design optimization methods can also provide a better means for system conceptual design of aerospace vehicles.

Table 5. Design results of HRM powered rocket.

Design variable	D_p (m)	e_1 (m)	n	F_i (kN)	p_{ci} (MPa)	α_i	ε	φ_0 (°)	Run time(min)
DDO	0.267	0.0322	4	14.3	1.89	2.66	4.6	85	499
KMCS	0.277	0.0316	4	14.31	1.93	2.79	4.5	85	1887
KTSA	0.277	0.0417	4	13.65	1.81	2.33	4.6	85	281

Table 6. Statistical results of HRM powered rocket.

Target/constraints	Method	μ	σ	Maximum	Minimum	μ/σ	G
M_R (kg)	DDO	333.97	2.8	342.26	326.31	0.00837	
	KMCS	353.36	2.89	362.62	344.92	0.00819	
	KTSA	364.41	2.93	372.95	356.37	0.00803	
Y_{max} (km)	DDO	119.28	4.02	131.18	107.51		0.429
	KMCS	127.04	4.23	139.47	114.85		0.954

	KTSA	128.59	4.56	141.94	115.89	0.978
	DDO	17.97	0.14	18.4	17.58	0.589
L/D	KMCS	17.5	0.14	17.94	17.12	1
	KTSA	17.57	0.13	17.92	17.22	1
	DDO	9.23	0.25	9.77	8.67	1
N_{xmax} (g)	KMCS	9.23	0.25	9.78	8.62	1
	KTSA	8.23	0.23	8.78	7.78	1

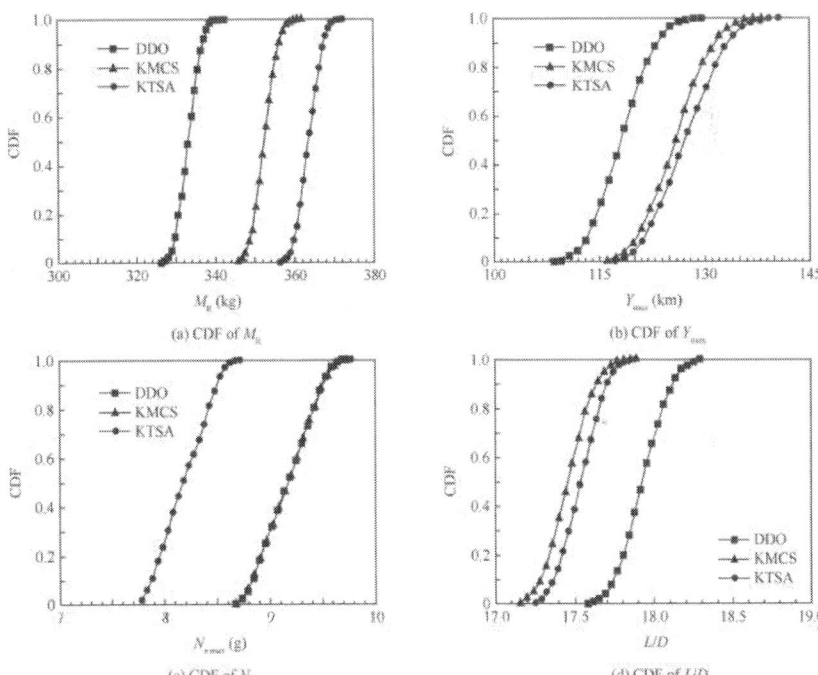

Figure 11. Cumulative distribution function (CDF) of the target and constraints.

CONCLUSIONS

(1) The multidisciplinary system design and analysis model of the HRM powered rocket is established and the input uncertain factors are quantified. The sensitivity analysis of the uncertain factors shows that among 42 uncertain factors the regression rate uncertainty has the

most significant effect on performances of the HRM powered rocket, thus it is necessary to accelerate the investigation on the combustion mechanism in HRMs.

(2) Two newly formed uncertainty analysis method including the KMCS and KTSA are carried out with a global approximation Kriging modeling method. The design optimization of the HRM powered rocket is carried out applying three methods including DDO, KMCS and KTSA. The results and comparisons show that the two UDO methods can provide design results with a higher reliability and robustness than DDO method and the KTSA method is more suitable for the initial design optimization phase while the KMCS method is more applicable for the detailed design optimization phase. The uncertainty design optimization methods can provide a better means for system concept design of aerospace vehicles.

(3) Due to the insufficiency of experiment or engineering data about the uncertain factors, there may be some inaccuracy on the related distributions and relative limit deviations. Our future work will focus on finding sufficient data to achieve accurate results. It is another important work to promote the engineering application of the UDO approach for HRM powered vehicles.

ACKNOWLEDGEMENTS

This work was supported by the National Natural Science Foundation of China (No.51305014) and China Postdoctoral Science Foundation (No. 2013M540842).

REFERENCES

1. Altman D. Hybrid rocket development history. Proceedings of 27th AIAA/SAE/ASME/ASEE joint propulsion conference; 1991 Jun 24–26; Sacramento. Reston: AIAA; 1991.

2. Davydenko NA, Gollender RG, Gubertov AM, Mironov VV, Volkov NN. Hybrid rocket engines: the benefits and prospects. Aerosp Sci Technol 2007;11(1):55–60.

3. Li XT, Tian H, Cai GB. Numerical analysis of fuel regression rate distribution characteristics in hybrid rocket motors with different fuel types. Sci Chin Technol Sci 2013;56(7):1807–17.

4. Casalino L, Pastrone D. Optimal design of hybrid rocket motors for microgravity platform. J Propul Power 2008;24(3):491–8.

5. Kosugi Y, Oyama A, Fujii K, Kanazaki M. Multidisciplinary and multi-objective design exploration methodology for conceptual design of a hybrid rocket. Proceedings of Infotech@Aerospace; 2011 Mar 29–31; St. Louis, Missouri. Reston: AIAA; 2011.

6. Dyer J, Doran E, Dunn Z, Lohner K, Bayart C, Sadhwani A, et al. Design and development of a 100 km nitrous oxide/paraffin hybrid rocket vehicle. Proceedings of 43rd AIAA/ASME/SAE/ASEE joint propulsion conference and exhibit; 2007 Jul 8–11; Cincinnati. Reston: AIAA; 2007.

7. Kniffen R J, McKinney B, Estey P. Hybrid rocket development at the American Rocket Company. Proceedings of 26th AIAA/ ASME/SAE/ASEE joint propulsion conference; 1990 Jul 16–18; Orlando. Reston: AIAA; 1990.

8. Cai GB, Zhu H, Rao DL, Tian H. Optimal design of hybrid rocket motor powered vehicle for suborbital flight. Aerosp Sci Technol 2013;25(1):114–24.

9. Rao DL, Cai GB, Zhu H, Tian H. Design and optimization of variable thrust hybrid rocket motors for sounding rockets. Sci Chin Technol Sci 2012;55(1):125–35.

10. Xuan Y, Xiang JH, Zhang WH, Zhang YL. Gradient-based Kriging approximate model and its application research to optimization design. Sci Chin Technol Sci 2009;52(4):1117–24.

11. Wang H, Wang SP, Mileta MT. Modified sequential Kriging optimization for multidisciplinary complex product simulation. Chin J Aeronaut 2010;23(5):616–22.

12. He LS. Ballistic missiles and launch vehicles design. Beijing: Beihang University Press; 2002.

13. Yao W, Chen XQ, Luo WC, van Tooren M, Guo J. Review of uncertainty-based multidisciplinary design optimization methods for aerospace vehicles. Prog Aerosp Sci 2011;47(6):450–79.

14. GJB 1026A–99, General specification for solid propellant rocket motors. Military standardization of People's Republic of China, 1999 Chinese.

15. GJB 2053A–2008, Specification for aluminum allot structural sheet for aerospace. Military standardization of People's Republic of China, 2008 Chinese.

16. Wen BC. Machine design handbook. Beijing: China Machine Press; 2010 Chinese.

17. Lucas JM. How to achieve a robust process using response surface methodology. J Qual Technol 1994;26(4):248–60.

18. Zhang RC, Wang ZJ. Design theory and data analysis of computer experiments. Chin J Appl Probab Stat 1994;10(4): 420–36 Chinese.

19. Xie YM. Research on robust optimization of sheet metal forming based on Kriging and grey relational analysisdissertation. Shanghai: Shanghai Jiao Tong University; 2007 [Chinese].

20. Rijpkema JJM, Etman LFP, Schoofs AJG. Use of design sensitivity information in response surface and Kriging metamodels. Optim Eng 2001;2(4):469–84.

21. Rao DL, Cai GB. Modified differential evolutionary algorithm for fast simulation optimization and its application. J Astronaut 2010;31(3):793–7 Chinese.

CITATION

Hao Zhu, Hui Tian, Guobiao Cai, Weimin Bao, Uncertainty analysis and design optimization of hybrid rocket motor powered vehicle for suborbital flight, Chinese Journal of Aeronautics, Volume 28, Issue 3, June 2015, Pages 676-686, ISSN 1000-9361, http://dx.doi.org/10.1016/j.cja.2015.04.015.

Index